THE DOCTORS' RIOT OF 1788

Lithographic print by T. C. Wilson believed to show anatomist William Hunter (right) with students performing dissections in his anatomy lab in London. "He was the first in great Britain that taught publickly dissections," said Hunter's brother, John, "for prior to this time, no pupil could get a subject, but what he could procure of himself, and when he ventured to get one, there was no one to instruct him." The print is based on a watercolor by Thomas Rowlandson, now in the collection of the Royal College of Surgeons of England. COURTESY OF THE HARVARD COUNTWAY LIBRARY.

THE
DOCTORS'
RIOT
OF 1788

BODY SNATCHING, BLOODLETTING, AND ANATOMY IN AMERICA

ANDY MCPHEE

PB Prometheus Books

Essex, Connecticut

Prometheus Books

An imprint of The Globe Pequot Publishing Group, Inc.
64 South Main St.
Essex, CT 06426
www.GlobePequot.com

British Library Cataloguing in Publication Information available

Library of Congress Cataloging-in-Publication Data Available

ISBN 978-1-4930-8805-8 (cloth)
ISBN 978-1-4930-8806-5 (ebook)

To Ma (both of them)

The body snatchers! they have come
And made a snatch at me;
It's very hard them kind of men
Won't let a body be!

Don't go to weep upon my grave,
And think that there I be;
They haven't left an atom there
Of my anatomie!

—Excerpted from "Mary's Ghost: A Pathetic Ballad"
by Thomas Hood

CONTENTS

Contents

AUTHOR'S NOTE

New York historian Robert Swan once described the warring historical accounts of the Doctors' Riot of 1788 as "incomprehensible." Said Swan, "The first full reports, which left many details wanting, surfaced in out-of-town newspapers two weeks later. Almost certainly newspaper editors witnessed this tragedy. Yet on Tuesday of the riot, Thomas Greenleaf, of the *Independent Journal and Patriotic Advertiser*, apologized to the public for being unable to give 'a concise statement of the sad confusion of the city since last Sunday for this day's paper.' Locally, any inkling of trouble could only be inferred two days later when physicians published notarized disclaimers of involvement with grave robbing of any church."

Swan was correct. Precious little contemporaneous information exists about the riot, making detailed examination of the event and its immediate aftermath difficult. However, four later accounts of the event offer enough citations and are written by such respected historians that they can provide a reliable and detailed history. One was written by James Joseph Walsh in his multi-volume work, *History of Medicine in New York: Three Centuries of Medical Progress*, published in 1919. Walsh was a devoted Catholic who earned his medical degree from the University of Pennsylvania in 1892 and was named dean of the medical school at Fordham University in 1906. Walsh also founded Fordham University Press, the oldest Catholic university press in the nation. His meticulous research of the riot, as well as the religious underpinnings of body snatching as a whole, provide a solid, scholarly base of information.

Another article on the event, by physician, author, and historian Jules Calvin Ladenheim, appeared in the Winter 1950 issue of the *Journal of the History of Medicine and Allied Sciences*. It, too, provides detailed and well-cited descriptions of the riot.

The third key account is Swan's own chronicle, "Prelude and Aftermath of the Doctors' Riot of 1788," which appeared in the October 2000 issue of *New York History*. Swan's piece offers a scholarly and dispassionate explanation of the riot, complete with references. Swan also refers to the fourth important resource, an article by Whitfield J. Bell Jr., a highly regarded authority in early American science and medicine. In the article Bell included a letter written April 16, 1788, from William Heth, a Virginia planter who served as a colonel in the Continental Army. He was writing to Edmund Randolph, a Virginia politician who succeeded Thomas Jefferson as Secretary of State. The letter, published in 1971, was discovered in storage at the Virginia State Library and contained previously unknown material about the riot.

I also examined and sometimes used information from Joel Headley's *The Great Riots of New York, 1712–1873*, Ralph G. Victor's "An Indictment for Grave Robbing at the Time of the 'Doctors' Riot' 1788," and from a marvelous speech given by William Alexander Duer to the St. Nicholas Society of New York in 1848. Those sources proffered great detail but few to no references. Thomas Gallagher's *The Doctor's Story* proved valuable in understanding the tenor of the times and the way doctors tended to work in the eighteenth century. How I wish Mr. Gallagher had provided at least a few citations for his detailed, highly embellished tales of the days and nights of the riot, but alas, he did not.

Using those sources and a few others, I tried to re-create a detailed account of the buildup to the riot, the riot itself, and the immediate aftermath. I admit to taking certain liberties in my account, however. For instance, I made several assumptions about the boys outside New York Hospital and their actions. Key sources indicate that "a boy" found a ladder and raised it to the anatomy lab window. If it was a single boy, he had to have been tall enough and strong enough to lift a wooden ladder, carry it to the window, and raise it up. A thirteen- or fourteen-year-old would fit that scenario, but it seems unlikely that a group of older boys would have been playing in the hospital yard even on a Sunday. They would have been working at home, perhaps, or already apprenticed. I decided, then, that the boys playing on the lawn were most likely eight to twelve years old and that at least two and probably three of them actually put the ladder up to the window. I could be wrong, but if I am, I don't think it's by much. In the interest of clarity, the reader should know, then, that those are the kinds of decisions I made in the chapters dealing specifically with the riot. Details in those chapters taken from historical accounts have been cited.

I decided to use terms we employ today that describe peoples of the time. For instance, rather than using "slaves," I used "enslaved Africans," the preferred use by officials of the African Burial Ground National Monument and Howard University researchers. I likewise used "American Indian," rather than older, more contentious terms, and I am aware that the great nations of the northeast ranged well into Canada, borders being nonexistent to indigenous peoples at that time.

The term "body snatching" occurs throughout, not "grave robbing." Grave robbing was reserved for people who stole personal items from a grave, items the theft of which could be prosecuted in court. Stealing a body, though, could not, seeing as how a body was not considered property. I also used "body parts" rather than "tissues" to make clear that although the term "tissue banks" sounds clean and rather clinical, it isn't. Although many tissue banks may be legitimate and operate ethically, many do not. So I wrote as if the body is made up of tissues, which it is, and that I should be clear about what those tissues actually are: body parts.

In most cases I have adjusted spellings and capitalization to more modern formats for easier reading. I did, however, maintain original spellings and capitalization where important for historical purposes.

It seems important at this point to let the reader know my own feelings about donating one's body for anatomical study, feelings that may or may not have played a role in writing this book, though I hope they haven't. I have had "Organ Donor" marked on my driver's license since my early twenties, and now that I am retired and moving (or perhaps creaking) through the last part of my life, I believe I will be donating my body to a university-based anatomy program when I'm finished with it. I know what can happen to my body after death. I know it could be handled harshly, disrespectfully, and, I suppose, amorally as well, and I accept that. I would be delighted even so to be used well after death, whether students studied my whole body or an orthopedic surgeon learned a new technique from my right shoulder, a military researcher discovered how to better protect soldiers from my torso, and a device manufacturer found a better way to reduce traumatic blood loss from my legs.

I accept those risks, but I also fervently wish that the federal government will finally push forward meaningful legislation to regulate body donations and thus prevent harsh, disrespectful, and amoral use of anyone's body. It's time, let's do it.

PREFACE

THE PHONE IN MY BEDROOM RANG LATE ONE NIGHT. IT WAS THE night shift supervisor at my hospital telling me that I needed to get to the operating room *stat*. I was an operating room technician then, what they call now a surgical technologist, and lived just a few minutes from the hospital, so I threw on my clothes and ran out the door.

When I reached the operating room, I immediately began setting up for an exploratory operation, having been told that someone in the emergency room was being brought to the operating room with severe chest and abdominal injuries. A young man on a motorcycle had gone off the road and crashed in a ditch. He hadn't been found for a few hours, and his injuries were life-threatening. The next thing I knew, two surgeons were bursting through the doors with a patient on a stretcher. The surgeons, gloveless and still in their street clothes, slashed open the man's abdomen, top to bottom. They thrust their bare hands into the man's belly, filled with blood at that point, and began searching. There was at least one artery bleeding profusely, and if they could find it and close it, maybe they could save the man's life.

I had helped with dozens of operations by then, Caesarean sections and gallbladder operations and hernia repairs, but I had never before been confronted by anything like that, and I will never forget it. The surgeons, two of the best I had ever worked with, rummaged around feverishly, desperate to find the bleeder, but it was no use. The patient died on the table from profound blood loss.

The surgeons, whom I respectfully and with profound gratitude call Roger and Jack here, were despondent but needed to know where the bleeding came from, so they kept looking. With blood no longer pouring into the abdomen, they soon found the culprit. "There it is," said Roger, the always calm voice in the room. The left iliac artery, an artery in the groin that feeds the leg, was splayed open about two inches, as if someone had pulled down its zipper. There was no way to recover from such an injury; it's a wonder the man had survived as long as he had. The two surgeons—one a grizzled veteran and the other fresh from his residency—began closing the cyclist's abdomen.

I've thought a great deal about that patient and that surgery while writing this book, about how absolutely critical anatomy knowledge is to physicians and about how, without it, they would be lost, as lost as that young man with a lacerated artery in a belly filled with blood. I've thought about the other organs I've had the opportunity to see and feel firsthand: the firmness of the edge of the liver, the white sheen of the diaphragm, the ruddy red of the non-pregnant uterus, and the stretched pink of a fully pregnant one. I've thought about some of the technological advances in anatomy education I've seen—video podcasts, virtual reality headsets, and anatomy tables that allow medical students to explore the body digitally in three dimensions. As brilliant, fascinating, and effective as those technologies might be, I'm not sure they can replace the kind of actual, hands-on-organs learning that working with cadavers can.

One of my challenges for this book, then, was to find a way to present the necessity of cadaver dissection hand in hand with the necessity of a rather despicable trade, that of the body snatcher, the resurrectionist, the dreaded sack-'em-up men. Those men were performing a vital service for medical education. They might have performed the service illegally, immorally, and purely for the money, but it was a service nonetheless. I can understand as well the fury that can come from someone desecrating the body of a loved one and particularly doing so without approval, without ever being notified of the act. But that looks at the issue through today's lens. At the time of the Doctors' Riot, society needed the resurrectionist's services just as it needs the services of morticians and pathologists today.

We still need people who purposely choose to deal with the dead and who treat our dead with kindness and respect. We need the dead, still, to help us learn. Even with the latest technologies, some of which are truly astonishing, we still need people who donate their bodies for study and practice.

The Doctors' Riot of 1788 sparked a movement that led to today's system of body donation and the laws and morality that support it. We have reached a place where whole-body donation has become not only accepted but is also seen as distinctly noble. We have reached a place where a wide variety of organs can be donated, removed from the body, and transplanted into someone else just prior to death. Yet while organ donation is closely regulated, the body donation business—and it *is* a business, to be sure—is almost completely unregulated.

The learning that occurs from practicing on real cadavers is just too damn important for our health. We need to protect that learning. I hope this book, in some small way, contributes to the creation of laws that provide a measure of assurance to families of people who have donated their bodies. We need to make sure that those bodies have not been used in vain and that they have not been manhandled, mistreated, neglected, or disrespected. We need, quite simply, to ensure dignity after death.

CHAPTER 1
MEDICINE'S MORAL CONUNDRUM

A WOODEN ENGRAVING FROM 1782 SHOWS A HARRIED-LOOKING man in wig and waistcoat standing in the middle of eight bodies resurrected on judgment day, among them a pair of one-legged men arguing about the ownership of a single severed leg, one man looking for his stomach and the other his head, and two skeletons greeting each other with fondness, one telling the other, "My dear Madam, I hope you are well. I am overjoyed to see you!" The man in the middle of this surreal cohort is the legendary British anatomist, obstetrician, and surgeon William Hunter, who cries out in anguish, "Never did I suppose that such a day would come!"

The caricaturist was scolding Hunter for using stolen cadavers for anatomy classes at a school he had opened in London's Covent Garden in 1746, foreseeing the day when his actions would come back, at least figuratively, to haunt him. Hunter had a long and distinguished career in London. He had been appointed physician extraordinary to Queen Charlotte in 1764 and as professor of anatomy at the Royal Academy in 1767, two preeminent titles, and was also a member of the Society of Antiquaries, the Royal Society of Medicine of Paris, and the Academy of Sciences of Paris. He grew wealthy enough to purchase an imposing house at 16 Great Windmill Street with enough room for an anatomy lab and a museum to house his growing collection of body parts. Students came to London from all over the world to learn from the great anatomist.

Hunter typically supplied students with their own cadaver, some of which he had obtained from Tyburn Gallows, about a mile from the school. Dozens of men, women, and children were hanged there every year for any of two hundred hangable offenses, including the theft of rabbits, blackening of the face at night, picking pockets, or cutting down trees. A writer in the Quaker journal *The British Friend* explained the plethora of crimes punishable by death at the time: "We hanged for everything—for a shilling, for five shillings, for forty shillings, for five pounds, for cutting down a sapling! We hanged for a sheep, for a horse, for cattle, for coining, for forgery, even for witchcraft—for things that were and things that cannot be."

Tyburn Gallows, however, couldn't supply all of the bodies Hunter needed for his students to dissect. He needed another avenue for obtaining cadavers and so turned to his own students or, more often, to resurrectionists, more commonly known as body snatchers or sack-'em-up men. These surreptitious sorts would closely monitor the comings and goings of gravediggers, particularly those working in the local potter's field or Black cemetery. After a burial, two or three sack-'em-up men would appear in the night and dig into a grave, open the coffin, sling a metal hook or other tool around the person's head or shoulders, and drag the body aboveground. The body would then be wrapped in a tarp for concealment and tossed into a carriage for transport to an anatomy lab, a doctor's office, or another place away from prying eyes. The robbers would be handed a fee for the package, and the body would be dissected in an anatomy class soon thereafter. Speed was essential in these matters; an unrefrigerated body begins to decompose within a day or two, making dissection and the learning of anatomy progressively more difficult.

The first known case of body snatching occurred in 1319 in Bologna, Italy, when four medical students ransacked a grave and stole the body inside. The brilliant artist Michelangelo Buonarroti was a teenager in the late 1400s when he began dissecting bodies to learn more about the structure of muscles and bones. Leonardo da Vinci paid body snatchers to provide cadavers to him, dissecting in the crypt of Santa Maria Nuova thirty cadavers between 1489 and 1513. Andreas Vesalius, the physician and anatomist, not only stole bodies from graveyards himself, but he also taught his students to do the same. Physicians throughout Europe depended on the theft of cadavers even though laws forbidding it were common. Stealing a body was actually not a crime. Ruth Richardson, in her seminal work, *Death, Dissection, and the Destitute*, explained: "Given the state of the law at the time, the dead human body did not constitute property, so taking a corpse from its grave was offensive, but not theft. The bodysnatchers could not fairly be called thieves on the ground that they 'stole' dead bodies."

Taking a body from a grave might not have been considered theft, but taking clothing or other articles buried with the body was. Laws for hundreds of years recognized those items as property and, therefore, subject to laws governing other types of property. Individuals who pilfered necklaces, rings, clothing, or any other articles buried with a person were subject to arrest and were known as grave robbers. Body snatchers, on the other hand, generally took great pains to leave those properties behind and take just the body.

Digging up a body was not a task for the weak, physically or morally. Resurrectionists tended to be shady characters, sneaky, often brutish, and, at times, startlingly clever. Without those brutes, though, prospective physicians were at a distinct disadvantage. Wrote one physician in 1824, "If dead bodies cannot be procured, it will be impossible for the pupils to learn anatomy, and without anatomy, neither surgeons nor physicians can practice with the least prospect of benefiting their patients."

A great many anatomy students entered the body snatching world as part of their education. They might not have enjoyed the task, but they knew that without a body to study, their education would be incomplete. Some students apparently did enjoy it, though, or at least took pride in their work. A group of students at Harvard College formed in 1770 a secret society dedicated to body snatching in the name of education. The group, known as the "Spunker Club," included in its membership future Massachusetts governor William Eustis; young Samuel Adams Jr., son of the famed politician of the same name; and John Warren, the founder of Harvard's medical school. Warren once told of a body snatching in late 1775 that "was done with so little decency and caution that the empty coffin was left exposed. It need scarcely be said that it could not have been the work of any of our friends of the Sp——r Club. It must have been the act of a reckless agent or a novice."

The importance of dissection for anatomy education, coupled with the difficulty in obtaining enough bodies for study, could imperil the education of even the brightest medical students. Take the case of a young man who had been turned away from the Royal College of Surgeons in London in 1828. According to famed British surgeon Sir Astley Cooper, the man had been rejected "on account of his ignorance of the parts of the body. It was found, on enquiry, that he was a most diligent student and that his ignorance arose entirely from his being unable to procure that which was necessary for carrying on this part of his education."

That "most diligent student" would have run into the same obstacle had he lived in the colonies. Few formally educated physicians existed in the colonies, and most of those had trained at colleges in Paris, London, or Edinburgh. Anatomy formed the core of that instruction and was taught largely by doctors and anatomists unaffiliated with any college or university. "The eighteenth century Anglo-American world had a strong tradition of independent anatomical schools," explained Stephen Novak, head of Archives and Special Collections at Columbia University. "The schools taught anatomy and surgery but didn't grant degrees. At a time when most American physicians gained their training via apprenticeships, an actual Doctor of Medicine degree wasn't absolutely necessary to practice medicine."

Starting in the latter half of the 1700s, the need for physicians in the colonies increased, as did the demand for cadavers. Doctors, anatomists, and sometimes even members of the lay public began offering courses on anatomy to anyone interested. Samuel Clossy, an Irish physician, taught anatomy and performed multiple dissections in

America in the mid-1760s, including one on a young White woman and one on a young Black man. Clossy had scheduled a third dissection but couldn't find a suitable body. He wrote, "We could not venture to meddle with a white subject, and a black or Mullato I could not procure." Physicians, then, were left with three choices, spelled out in clear detail by author and journalist Thomas Gallagher: "They could abandon the teaching of anatomy altogether and thereby stunt the growth of American surgery, they could teach theory and send their students out to practice on live patients, or they could obtain surreptitiously what they needed in order to teach the subject as every European medical school agreed it should be taught."

Without enough bodies for dissection medical students could learn anatomy only from illustrations in lectures or in books, and not until the mid-1800s would a complete and accurate anatomy textbook become available. Ann Zumwalt, associate professor of anatomy and neuroanatomy at Boston University School of Medicine, explained the importance of the kind of hands-on learning available only on cadavers. "It's a fundamental shift in awareness," she said. "Students go in, first day of anatomy, and they're thinking, *This looks like a mess to me. I don't understand what I'm looking at.* Pretty quickly, though, they think, *Oh,* this *is what a nerve feels like,* as opposed to an artery, even though they kind of look alike. *I can tell by feeling what the difference is.*" How, then, could medical students in the late 1700s have gained a three-dimensional understanding of the human body without dissection?

Therein lies a moral conundrum that has confounded medicine since at least the Middle Ages: Is it moral to dissect a body for the betterment of medical students if the body was illegally obtained? Said another way: Does the future health of the living outweigh society's need to maintain the dignity of the dead? Three days of deadly protests in post-Revolutionary New York City brought those questions into stark relief. The protests would end up being known as the Doctors' Riot of 1788, not because physicians themselves rioted but because the public rose up in outrage against physicians and anatomy students. Most bodies to that point were stolen from the pauper's cemetery and the Negroes Burying Ground, today referred to as the African Burial Ground National Monument. At some point, however, students began taking bodies from the cemetery of nearby Trinity Church, a church mostly for Whites.

Several stories exist about how the riot started, the most common, but not necessarily correct, being that on Sunday, April 13, 1788, a group of boys were playing outside an anatomy lab at New York Hospital. A student named John Hicks Jr. noticed the boys and decided to hang a severed arm out of the window. "This is your mother's arm!" Hicks told one boy. "I just dug it up!" As it happened, the boy's mother had, in fact, just died. Stricken, the boy ran home to tell his father of the gruesome display. The father immediately set out for his wife's grave and found, to his horror, that her grave had been disturbed and her body gone. The theft was too much to bear for the aggrieved man, and he set about gathering some friends and coworkers to storm the hospital.

The group swelled over the next two days into a mob that searched the city for Hicks and his fellow anatomy students, as well as their instructors, to give them a good thrashing. Some medical men escaped immediately; others were rescued and taken to the prison for protection. At one point the great Alexander Hamilton is said to have stood on the steps of a medical building at Columbia College and tried, unsuccessfully, to calm the growing crowd.

The mob eventually found its way to the prison, where it met the city's mayor, James Duane, and trained militia defending the students. Alongside Duane were John Jay, one of the Founding Fathers and the nation's first Chief Justice of the Supreme Court, and Revolutionary War hero Baron von Steuben, George Washington's Prussian aide-de-camp who turned a ragtag army at Valley Forge into a disciplined fighting force. They all tried to quell the crowd's anger but were no match for the fury thrown their way. Finally, rioters pelted the militia and attendant dignitaries with stones, dirt, clubs, and brickbats. Jay and Steuben were each struck in the head. The mayor at last ordered his militia to fire into the crowd, and in the melee that followed, several rioters were killed and many wounded. The mob dispersed and never gathered again.

New York soon passed laws to prevent body snatching, as did a number of other states, but the laws proved more cosmetic than practical. Between April 1788 and the end of 1884, a total of twenty-two riots broke out against body snatching, with three in 1788 alone: the Doctors' Riot in New York, a second riot in Baltimore, and a third in Boston. Riots broke out in Zanesville, Ohio, in 1811, New Haven in 1824, and Saint Louis in 1844. Berkshire Medical Institution in Pittsfield, Massachusetts, suffered two riots: one in 1830 and another a decade later. Castleton Medical College in Vermont nearly suffered a riot in 1830. The college had decreed in 1824 that no bodies would be taken from Vermont cemeteries for the study of anatomy and that they would all come from cities in neighboring states.

"It was usual for all medical colleges to have rules forbidding dissection," wrote anatomist Arthur M. Lassek in his book *Human Dissection: Its Drama and Struggle*, "but they weren't necessarily observed. They were made primarily to mislead the public." So when some townspeople in nearby Hubbardton discovered the grave of Mrs. Penfield Churchill empty, a swarm of three hundred citizens stormed the school. A search led to the discovery of a headless body hidden behind the wainscoting. The townsfolk agreed not to press charges against the school or its students if they returned the unfortunate woman's head. During the discussion, a student snuck out of the room and returned a short time later with the head, which had been hidden in a barn. No arrests were made, and both sections of Mrs. Churchill were reburied.

The incidence of body snatching began declining during the Civil War, when embalming practices became, out of necessity, more refined. As embalming improved, especially after the discovery of formaldehyde in 1869, bodies could be preserved longer, obviating the need for fresh bodies to dissect. Medical education made enormous improvements during the nineteenth and twentieth centuries. The apprenticeship model disappeared, and colleges required a greater number of courses for graduation. Anatomy was still required, but so too were courses on pathology and diagnosis, and new, more specialized courses appeared, from growth and development to histology and endocrinology. As medical education improved, so too did the public's perception of medical education, and along with that came a greater willingness of citizens to donate their own bodies to science.

Without a commercial need for body snatching, the trade of resurrectionists gradually died out, though not completely. In 1902, a ring of body snatchers was discovered in Indianapolis, Indiana. Twenty-five people were indicted, including an undertaker's assistant named Rufus Cantrell and an anatomy teacher at Central College of Physicians and Surgeons named Joseph C. Alexander. An article in the *Medical Standard* that year condemned the men:

> *Grave robbing, naturally enough, is regarded with especial horror and detestation, and professional men must look upon it with small measure of satisfaction or approval. It was doubtless excusable, perhaps necessary, a century ago—even less—when the law made no provision for anatomical material, so essential to correct instruction in medicine. But there should be no excuse for it now.*

Although the traditional form of body snatching faded away, a more insidious version appeared in its place—bodies willed to a medical school or tissue bank being surreptitiously dismembered and sold in a market many people even today don't know exists. Anatomy courses at medical schools are the most obvious and socially accepted market for willed bodies, but others are not obvious and much less accepted, including medical device manufacturers, for-profit surgical training centers, automakers, and US military research facilities. Medical device manufacturers need actual human bodies to develop technologically advanced instruments for surgeons. Dentists and surgeons need to practice new techniques on the dead before working on the living. Automakers and the military need to know how certain forms of trauma actually occur in humans involved in a car crash or a bombing. Ugly but necessary.

Most medical ethicists urge anyone considering donating their body to be fully aware of all the possible ways their body might be used. Unfortunately, that doesn't happen often enough. Too many bodies today are being used in ways the donor never expected or wanted. Perhaps the most famous victim of modern-day body snatching was the late British author and journalist Alistair Cooke, renowned for hosting the PBS show *Masterpiece Theatre*. A body snatching ring led by a former Brooklyn dentist named Michael Mastromarino removed Cooke's bones after his death in 2005 and replaced them with PVC pipes to provide the proper body shape for viewing in an open casket. The bones were then sold for more than $7,000 to a legitimate company for processing into dental implant material. As it turned out, Cooke was one of a great many of Mastromarino's victims in his four-year-long grave-robbing spree. The case became the subject of a 2010 documentary called *Bodysnatchers of New York*.

In another disgusting case, an embalmer from Staten Island was sentenced in 2009 to as many as twenty-four years in prison for his role in a scheme to illegally strip and sell skin, bone, and other body parts from more than a thousand bodies between 2001 and 2005. One more case: In the summer of 2023, a former morgue employee at Harvard Medical School and his wife were arrested for the theft and subsequent sale of body parts donated to the school.

Those types of incidents aside, dissection remains a core element of the education of medical students and students in other caring professions, such as dentistry, physical therapy, and mortuary science. In a medical school, teams of students are assigned an individual whose body had been donated to the school. The donor serves as the student's first and, in many ways, most important patient. The students generally form a close relationship with their donor, one that can last a lifetime. Sometimes that relationship starts with a single spellbinding encounter. An anatomy instructor at the Ohio

State College of Medicine recalls one of those moments. "I remember walking by a dissection table," said the instructor, "only to notice a student speechlessly gazing down at a brain. As I approached the table, she suddenly looked up at me, her eyes filled with rapture and awe, and said 'This is so amazing! This structure once controlled a person's entire life—every thought that they ever had, every movement that they ever made, every emotion that they ever felt.'"

Mark Norton, a former medical student at Georgetown University, talked at his graduation about the contribution of donors to his own education. "They knew nothing about us," said Norton, "and yet they dedicated their final act on this Earth to share their most intimate possession with us in the hope that we could learn from them. Our donors taught us to celebrate life and to never forget the need for humanity and compassion in medicine—a lesson that could never be explained in any textbook or on any app."

CHAPTER 2

A NATION AT PEACE AND IN TURMOIL

THE EMINENT BENJAMIN FRANKLIN IS HERE, SITTING AT THE HEAD of a table. He is wrapped in a dark frock, his long wavy hair falling to his shoulders, his mouth a thin crease above a slightly clefted chin. His dark, aged eyes peer at the viewer in kindness and perhaps fatigue as well. John Adams is here too, his head adorned in a powdered wig, his face as impassive as a napping sloth. Behind Adams stands a slender figure in a brown frock and tan waistcoat. He has sizable ears and a conspicuous nose. His right hand holds a rolled document; his left points toward Franklin. He stands erect, seemingly secure in his position at this most historic event.

The man is the remarkable John Jay, coauthor of the *Federalist Papers*, first Chief Justice of the US Supreme Court, and, later in his storied career, Governor of New York. Politics ran deep in Jay. His great-grandfather, Jacobus Van Cortlandt, had served two terms as mayor of New York City in the early 1700s and had held a number of judgeships there as well. Jay himself helped draft the state's first Constitution, served in the Continental Congress, and, as the nation's ambassador to Spain during the Revolutionary War, tried desperately to obtain war funding, a feat that eventually proved impossible. Perhaps more important, he led the negotiations to develop the treaty that officially ended the war.

Jay was thus placed centrally, with Adams and Franklin, in a painting formally called "American Commissioners of the Preliminary Peace Negotiations with Great Britain," more commonly known as the "Treaty of Paris." The painting was the concept of

Benjamin West's unfinished painting of the preliminary Treaty of Paris. The final treaty, signed in 1783, was essentially the same as the preliminary treaty of November 1782. From left: John Jay, John Adams, Benjamin Franklin, Henry Laurens (Founding Father and former president of the Continental Congress), and William Temple Franklin, Benjamin's grandson. GIFT OF HENRY FRANCIS DU PONT. COURTESY WINTERTHUR MUSEUM, GARDEN & LIBRARY.

Benjamin West, an American portraitist who had moved to England as a young man and had quickly risen to the top of the British art world, eventually becoming Historical Painter to the King and a founder and second president of England's Royal Academy of Arts. West traveled from London to Paris in late 1782 to illustrate the signing of the historic treaty and chose to place American delegates at one end of a table and British delegates at the other. He planned to donate the painting to Congress.

"This work I mean to do at my own expense," West explained, "and to employ the first engravers in Europe to carry them into execution, not having the least doubt [that] all will be interested in seeing the event so portraid." The final painting was to include each nation's negotiators: Franklin, Adams, Jay, and fellow Founder Henry Laurens for the Americans, and a Scottish merchant named Richard Oswald for the British. Laurens, however, had been captured on his way to Paris and jailed in the Tower of London. He was too ill on his release to travel to Paris and take part in the remaining negotiations. Oswald wasn't available to sit for West either, having been forced to resign his position after the preliminary negotiations. Without Laurens or Oswald available to sit for the portrait, West set the work aside and went on with his life. Adams felt hurt but hopeful that the great artist would one day finish the painting. "As I very strongly expressed my regret," said Adams, "that this picture should be left unfinished, Mr. West said he thought he could finish it." West never returned to the painting.

Here was one of the most important artists of the day, a man who had painted portraits of King George III, who had trained John Trumbull, famed painter of the American Revolution, and who was known to the cognoscenti of central London as the "American Raphael," now forced to leave a painting of five of the most important men in the world at a monumental event in history half-finished. Unthinkable. Yet if ever a piece of art captured the essence of post-revolutionary America and its relationship with its former overlord, this was surely it.

Just as Jay, Franklin, and Adams formed the foundational pieces of the painting, so too were the foundational pieces for the new republic put in place by war's end. They were men whose names resonated with honor and respect then and today: George Washington, Alexander Hamilton, John Adams, Thomas Jefferson, James Madison, James Monroe, and, of course, the elder statesman Franklin. As these and other American leaders emerged, English leaders faded into the background.

Washington had already grown into his role as commander of the military and continued to mature into his role as commander in chief. The men who would fill other leadership positions in the new government had also demonstrated their administrative mettle.

Adams, Jefferson, Madison, and Monroe, the youngest of the group, would all serve as president during the nascent years of the nation. New York's George Clinton, along with the infamous Aaron Burr, would each serve as vice president and leader of the US Senate. Frederick Muhlenberg, a Lutheran minister and skilled politician from Pennsylvania, was elected as the nation's first Speaker of the House, and Washington's secretary and aide-de-camp during the war and renowned Connecticut politician, Jonathan Trumbull Jr., served as the nation's second Speaker.

As competent as those leaders surely were, they faced governing under extraordinarily difficult circumstances. Local, state, and federal coffers were in shambles, with massive war debts plaguing the nation. Laws and regulations at all levels were being created, rewritten, or eliminated. Until the Constitution was ratified in June 1788, the government lacked a set of overarching principles for resolving complex disputes. It had no power to extract taxes from individual states nor the money to maintain the military. No federal judiciary existed to rule on conflicts between states. No central bank existed to issue standard currency. All of those issues and more needed to be overcome for the new nation to function capably in a changing world order.

At the same time, a still-rebellious national citizenry was turning against Loyalists, people who had demonstrated loyalty to the Crown, irrespective of race, religion, or nationality. Oscar Zeichner, professor emeritus of American history at City College of New York, explained that "Patriots with real or fancied grievances were not in a mood when the war ended to listen sympathetically to arguments urging the desirability and justice of forgetting the past. Most of them were determined not to permit the Loyalists to reside in or return to the state." So many instances of violence against Loyalists occurred during the summer of 1783 that Tories throughout the colonies fled to New York, where British forces still present could offer them some protection. Violence broke out north to south. "There was a lot of bloodshed," said Maya Jasanoff, history professor at Harvard University, "particularly in the South. Gangs of revolutionaries and gangs of Loyalists would attack each other, go to each other's plantations. In fact, some of the big battles in the South happened after the surrender at Yorktown."

Throughout the colonies, Patriots harassed Loyalists, seizing their homes and tarring and feathering them for their support of the Crown. By 1783, some 60,000–80,000 Loyalists of all shades had fled American shores. The remaining Loyalists were left to adapt to their new circumstances: the vanquished surrounded by the victors. Should the vanquished be allowed to live freely in the new nation, or should they lose their citizenship and rights? Should they be punished for their stance, or should the nation turn its other cheek? Those questions hammered at the root of America's transition from British rule to self-rule, from a cluster of colonies laboring under a monarch's heavy thumb to a United States of America suddenly responsible for its own survival.

Loyalty to the Crown meant something else entirely for Black citizens, free and bound alike. When the war began, the colonies were home to more than 500,000 Blacks, most of them enslaved. The Continental Congress initially decided that its army should exclude Blacks, a decision that proved shortsighted and would be countermanded when regiments couldn't fill their enlistment quotas with White soldiers. Congress's snub of such a large number of potential soldiers reverberated throughout the colonies and gave a great many Blacks pause when deciding which side to support.

Britain took full advantage of that snub and opened its own recruitment efforts for Blacks. Virginia governor Lord Dunmore issued a proclamation in 1775 promising freedom to defecting slaves, a decision based not on moral grounds but on practical ones. Britain planned to use escaped slaves as construction and maintenance workers, which its army desperately needed. "New York City, occupied by the British for seven years during the war, became a mecca for fleeing New York slaves," explained David Kobrin in *The Black Minority in Early New York*. "Most of the former slaves who made it to the British zone were hired as paid laborers on military works in and around the port city; but some Negroes were allowed to join Loyalist fighting units under British control."

As many as 20,000 free and enslaved Blacks would eventually join the Crown's cause. It seemed clear to them that England would eventually free them, or at least protect them, and that an independent America would probably continue to enslave them. After the British surrender at Yorktown in 1781, thousands of Black Loyalists jumped aboard British ships bound for England, Nova Scotia, or the Caribbean. British historian Simon Schama explained that "in return for their loyal service [to the Crown] in the late American war, they were to be granted two gifts of unimaginably precious worth: their freedom and their acres."

Those acres, though, were mostly located in inhospitable places. In Nova Scotia, for instance, Black Loyalists were granted land mostly in rocky, inarable areas. "The blacks had no way, most of them, to clear and work [the land]," wrote Shama, "unless they hired themselves or their families out to the white Loyalists. That meant more cooking and laundering; more waiting on tables and shaving pink chins; more hammering rocks for roads and bridges. And still they were in debt, so grievously that some complained their liberty was no true liberty at all but just another kind of slavery in all but name."

American Indian nations that had supported Britain, as well as those that had supported the Patriot cause, suffered equally. Centuries before Europeans came to America, an alliance had formed among six American Indian nations in New York: the Cayuga, Mohawk, Oneida, Onondaga, Seneca, and, later, the Tuscarora nations. English-speaking people called the alliance the Iroquois Confederacy. Indigenous peoples called it Haudenosaunee (pronounced hohden-oh-SHAW-nee), which means "People of the Long House" or "People Who Build Houses." Haudenosaunee leaders initially tried to stay out of the war, siding with neither the Patriots nor the British. The chiefs soon realized, however, that neutrality was untenable, and the confederacy broke apart. The Mohawk, Onondaga, Cayuga, and

Seneca nations threw their weight behind the British; the Oneida and Tuscarora nations chose to support the colonists.

In the end, it mattered not which side the nations had supported. No American Indian leader was present at or invited to treaty negotiations in Paris, nor did the final Treaty of Paris delineate what would become of any American Indian nation after the war. With no legal protection, American Indians and their lands became prey to settlers pushing west. From the mid-1780s until the end of the Civil War, White settlers, backed by the US Army, pushed westward. Untold numbers of American Indian men, women, and children were slaughtered and their land confiscated. "The war left their confederacy and culture shattered," wrote William Sawyer for the National Parks System, "and their lands and villages devastated and destroyed. While time and fortune has helped, many wounds from that time have yet to heal."

A form of legalized revenge against Loyalists began well before the war ended, with colonies passing laws that allowed their governments to confiscate Tory-owned land. New York, in particular, proved strongly anti-Loyalist and passed several draconian laws to make Tories pay dearly for supporting the Crown. The Forfeiture Act of 1779, for instance, was used to confiscate the land of Loyalists and banish them from the state "on pain of death." The act specified that the oath of a single "credible" witness was enough for a grand jury to indict any alleged Loyalist. Similar laws allowed New York to tax Loyalists more heavily than other citizens and banish Loyalists from their own homes. One such Loyalist, David Colden, the son of a former lieutenant-governor of New York, had been forced to leave his Long Island home in 1783 by Whigs intent on ridding the state of Tories. "Cursed, cursed tyrants," wrote Colden, "who drive me from my wife and children and put it out of my power to assist or comfort them when they need it most."

Historians generally consider the anti-Loyalist laws of New York harsher than those of every other state. Joseph Tiedemann, emeritus

professor of history at Loyola Marymount University, explained
New York's intolerant treatment of Loyalists this way:

> *This hostility was in part a result of the bitter, protracted civil war
> that had so recently divided New York, but the laws also expressed the
> patriot's need for psychic reassurance that their efforts had not been in
> vain and that they could secure in peace what they had won in war.
> However, once the state had made explicit the kind of peace the more
> vengeful would impose if Royalists refused to accept the new political
> order, other Patriots began working for a compromise solution by advo-
> cating leniency.*

John Jay sided with those supporting leniency, but only to a point.
He believed that condemning or expelling Loyalists was too harsh
a response and that all Tories "except the faithless and cruel" should
be excused. He railed against more extreme views and urged "clem-
ency, moderation, and benevolence." It was Jay's friend, Alexander
Hamilton, however, who became the most consistent and outspoken
proponent of conciliation and moderation in dealing with Tories.
Hamilton's arguments for conciliation were based partly on international law and partly on his own core beliefs about justice, arguments that would serve him well, not only as the primary author of
the *Federalist Papers* but also as the principle defender of persecuted
Tories in New York.

Hamilton had, by war's end, become a familiar face in the city,
a well-known revolutionary in his mid-twenties who cut a glamor-
ous swathe through high society. Slim, upright, and handsome, with
long, reddish-brown hair braided behind and tied with a black rib-
bon, the young man exuded elegance, grace, and style. He passed
the bar exam in 1782 and found almost immediate success arguing
cases in court. Hamilton had been blessed with a mellifluous voice
and could speak eloquently and at great length on even mundane
matters. Robert Troup, Hamilton's colleague and former roommate
at the College of New Jersey (now Princeton), recalled that Hamilton sometimes talked altogether too much. "I used to tell him," said

Troup, "that he was not content with knocking [his opponent] in the head, but that he persisted until he had banished every little insect that buzzed around his ears."

Hamilton's persistence paid off when arguing at court for Loyalist clients who had been sued under the state's anti-Tory laws. Most of those clients were moneyed landowners. Hamilton, perhaps more than most other lawyers, realized that Loyalist wealth would ultimately benefit the new nation in myriad ways. He believed that individuals previously sworn to the Crown could and should be reintegrated into American society. "The full truth of Hamilton's motivation for defending Loyalists is complex," wrote Hamilton biographer Ron Chernow. "He thought America's character would be defined by how it treated its vanquished enemies, and he wanted to graduate from bitter wartime grievances to the forgiving posture of peace."

Hamilton grew particularly annoyed at New York's Trespass Act of 1783, which allowed Patriots who had owned property confiscated during the war to sue Tories who had seized or damaged that property. The Trespass Act was based on intense feelings of revenge against Tories for their wartime actions. In New York City, those feelings stemmed in part from the cruelty dispensed on captured citizen-Patriots. More than 11,000 Americans captured during the war died aboard the sixteen or more British prison ships harbored in Wallabout Bay on the East River, succumbing to barbaric conditions and unspeakable cruelty. So many prisoners died each day on one prison ship, the notorious HMS *Jersey*, that the guards took to lifting the hatches each morning and yelling out, "Up on deck, you damned Yankee Rebels, and turn out your dead!"

New Yorkers demanded retribution, and three anti-Loyalist acts—commonly called the Trespass, Forfeiture, and Citation acts—had been designed to provide just that. One writer claimed that George Clinton, New York's governor from 1777 to 1795, "had rather roast in hell to all eternity than . . . shew mercy to a damned Tory." Hamilton took a broader view, believing the acts would harm the nation and that they countermanded the state's own constitution. Writing under the pen name Phocion, he wrote that "the 13th article of the constitution declares, 'that no member of this state shall be disfranchised or defrauded of any of the rights or privileges sacred to the subjects of this state by the constitution, unless by the law of the land or the judgment of his peers.'" He insisted that the state legislature "could not, and cannot, without tyranny, disfranchise or punish whole classes of citizens . . . without trial and conviction." Hamilton further asserted that the acts stood in direct contradiction to the central tenets of the Treaty of Paris, particularly Article Six, which stipulated that Loyalists shall not "hereafter suffer in life or person, or be deprived of his property, for the part he may have taken [during the war]." Hamilton believed that the Trespass Act, an act of a single state's legislature, must not take precedence over a peace treaty signed by the federal government. It was an argument he would invoke over and over in defending Tory clients.

Other defense lawyers knew, as Hamilton did, that the Treaty of Paris could be used to oppose charges brought under the Trespass Act, but they consistently chose not to present such a defense. "The fact is," wrote Hamilton in a letter to George Washington, "that from the very express terms of the Act a general opinion was entertained embracing almost our whole bar as well as the public that it was useless to attempt a defence—and accordingly . . . many compromises were made and large sums paid under the despair of a successful defence. I was for a long time the only practicer who pursued a different course and opposed the Treaty to the Act."

In perhaps the most famous Trespass Act case, *Rutgers v. Waddington*, Hamilton "elaborately and learnedly" argued on behalf of Joshua Waddington, a Loyalist who had been assigned by the British Army to manage a brewery and alehouse during the war. Before the war, the brewery had been owned and operated by Elizabeth Rutgers, a widow who had assumed ownership when her husband died. When the British took over the city in 1776, Rutgers and her sons fled the property, though not before stripping it "of everything

John Trumbull's famous portrait of the remarkable Alexander Hamilton. COURTESY OF THE NATIONAL PORTRAIT GALLERY, SMITHSONIAN INSTITUTION; GIFT OF HENRY CABOT LODGE.

of any value except an old copper [vessel], two old pumps, and a leaden cistern full of holes." Waddington, as the new brewery manager, was ordered by the British Army to refurbish the facility, which he did at a personal cost of £700. Waddington also paid rent to the army from the time he took over until late November 1783, when a fire destroyed the brewery.

After the war ended, Rutgers returned to New York City and took back control of the house and pub, demanding compensation from Waddington. When Waddington refused, Rutgers filed suit in the Mayor's Court, over which presided the mayor and five city aldermen. She asked the court for a judgment under the Trespass Act of £8,000, an amount she claimed covered Waddington's rental of her property, ignoring the cost borne by Waddington for renovating the facility and the rent he had paid the British throughout his time there.

Both Rutgers and Waddington were represented by lawyers with superb credentials and fine legal brains. Rutgers hired attorney John Laurance, who would a year later become a delegate to the Continental Congress, and Robert Troup, Hamilton's friend and former roommate. Her team also included New York attorney general Egbert Benson, a man "more distinguished than any man among us," according to James Kent, a clerk for Benson during the trial, "Hamilton alone excepted." Benson also happened to be Rutgers's nephew. Having a relative serve as counsel would be unthinkable today but was relatively common at the time.

Waddington was represented by Hamilton and two esteemed co-counsels, Brockholst Livingston and Morgan Lewis. Livingston was a classmate of James Madison and, in 1779, served as John Jay's private secretary during a diplomatic mission to Spain. He was later appointed by President Thomas Jefferson to the Supreme Court. His partner, Lewis, began working in Jay's law office as soon as he graduated from the College of New Jersey and was later named to New York's Supreme Court. In 1804 he bested Aaron Burr in a campaign for governor of New York.

The case came before the Mayor's Court on February 29, 1784, with fifty-one-year-old James Duane presiding, along with five city aldermen, all attorneys. Duane was a respected jurist who had served on the First and Second Continental Congresses and had only recently been appointed mayor. John Adams once described Duane as a man with "a sly, surveying eye, a little squint-eyed" and "very sensible, I think, and very artful." Counsels for Rutgers and Waddington presented their cases logically and persuasively, none more so than young Hamilton. Spectators in the courtroom "listened [to Hamilton] with admiration," recalled Kent, "for his impassioned eloquence."

Six months later, Duane released the court's verdict, which sided primarily with the defense and declared that Waddington owed no rent to Rutgers for the period he had been paying rent to the British Army, though he did owe for the period prior to British takeover. Peter Hoffer, a historian at the University of Georgia, explained the essence of the court's decision. "Duane suggested," said Hoffer, "that no state could unilaterally nullify a provision of the treaty ending the war, for that would be 'contrary to the very nature of the confederacy.'" With that decision, the Trespass, Forfeiture, and Citation acts fell one after the other.

Of far greater import, however, was the court's conclusion that if a state law conflicts with a treaty of the US government, the state's law must be ignored. The *Rutgers v. Waddington* decision proved monumental for a time when the nation possessed no federal judiciary, and it continues to stand today as a foundational principle defining state and federal jurisdictions. The case further set a precedent for the doctrine of judicial review, which holds that courts must make decisions according to the principles of national laws and that laws of individual states must also conform in principle to those laws.

Hamilton's success in the Rutgers case, for which he was paid £7, led to his pleading more than sixty similar cases in the coming years. Those cases provided Hamilton with a lucrative practice and launched his star even higher into the rarified world of celebrity than it had been when the war ended. He had, after all, worked with Baron von Steuben in Valley Forge, been Washington's right-hand man throughout much of the war, and had even commanded troops in the war's final contest, the Battle of Yorktown.

His state's eventual recognition of the unfairness of its anti-Tory statutes was emblematic of a nation trying to find itself, to become what its revolution had promised, to convert from a land of British citizens to a land of Americans. That conversion occurred gradually and proved anything but smooth.

CHAPTER 3
LIFE IN THE NEW REPUBLIC

FROM THE TIME KING JAMES I GRANTED A CHARTER TO ESTABLISH colonies in the New World to the time the colonies announced their intention to become independent, colonies operated according to British laws and customs. They based their currencies on the British system, their laws on the common law system operating in Britain, their government structure on the British parliament, their language on the King's language, and an infinite number of everyday matters on how the British people handled them. Roughly six generations of colonists had come of age as British citizens, protected and policed by officers of the Crown and restricted in their rights by a monarch thirty-five hundred miles away.

After the war, the nation found itself deyolked from the King but still economically dependent on trade with England and culturally enmeshed with British traditions. Kariann Akemi Yokota, in her illuminating book *Unbecoming British*, explained the incongruity: "The process of creating a separate society out of a people who so recently thought of themselves as British in outlook and tradition was fraught with hypocrisy and confusion. Americans vacillated between celebrating their future as a sovereign nation and struggling to overcome the cultural insecurity born of their colonial past." England did little to foster security in their former holding. Britain treated America like a naughty child, as did France. "Their words were sometimes those of age to youth," wrote historian C. Vann Woodward, "parent to offspring, master to apprentice, teacher to pupil."

In unbecoming British, America sought to become the next Europe, the next great cultural center of the world from which the greatest poets, artists, philosophers, and physicians would come. "Colonial Americans," wrote historian Joseph Ellis, "were profoundly aware that they lived on the periphery of a civilization whose center was London. Pre-revolutionary America was a provincial society whose leading members aped the manners of the English aristocracy and whose past accomplishments in the arts were derivative gestures, copies rather than originals."

Ben Franklin expressed his own jealousy about England's cultural accomplishments, writing to a friend in 1763, "Why should that pretty island, which compared to America is but like a stepping stone in a brook, scarce enough of it above water to keep one's shoes dry; why, I say, should that little island enjoy in almost every neighbourhood more sensible, virtuous, and elegant minds than we can collect in ranging a hundred leagues of our vast forests?"

The changeover from King's College, one of that pretty island's institutions in New York City, to Columbia College, an American institution, attests to the kinds of pervasive changes the nation needed to make to come into its own. King's College was established "for the instruction and education of youth in the learned languages, and liberal arts and sciences." First opened in a vestry room at Trinity Church, a temporary home until a college could be built, classes at King's started July 17, 1754, and consisted of eight students being taught thirteen subjects by Samuel Johnson, a respected Anglican minister from Connecticut.

Johnson proclaimed that although the college would be operated under the auspices of the Church of England, it would hold to the central tenets of the religions already established in America, among them Lutheran, Congregational, Quaker, and Presbyterian. Johnson said the college's goal was to "teach and engage the children to know God in Jesus Christ and to love and serve Him in all sobriety, godliness, and righteousness of life, with a perfect heart, and a willing mind; and to train them up in all virtuous habits and all such useful knowledge as may render them creditable to their families and friends, ornaments to their country, and useful to the public weal in their generations."

The college moved about two years later to, as Ron Chernow wrote in his book *Alexander Hamilton*, a "stately three-story building with a cupola that commanded a superb view of the Hudson River across a low, rambling meadow." The site today is ringed roughly by Barclay, Church, Murray, and West Broadway Streets. The building and its grounds were encircled by a fence, with its front gate guarded at all times by a porter, the better to keep students in and prostitutes roaming the adjacent red-light district out.

The college soon graduated several pivotal figures in the nation's founding, including John Jay, who graduated in 1764; Robert R. Livingston, who graduated in 1765 and served on the committee to write the Declaration of Independence; Gouverneur Morris, who graduated in 1768 and later wrote the Preamble to the Constitution; and Alexander Hamilton, who graduated in 1778 and had, even as a student, been proving his worth to the nation, playing a central role in persuading the Marquis de Lafayette to send a fleet of warships to the revolutionaries.

King's opened a medical school in 1767, motivated in part by professional jealousy. The school's founding faculty member was a twenty-five-year-old physician named Samuel Bard. Born in Phila-delphia and trained in Edinburgh, where many colonial physicians at the time gained their training, Bard had received his predilection toward medicine most likely from his father, John Bard, a highly regarded physician in his own right and a good friend of Benjamin Franklin. Samuel Bard had heard while studying in Edinburgh that Philadelphia was about to open a medical school, the first ever in the colonies. "I feel a little jealous of the Philadelphians," Samuel wrote to his father in 1762, "and should be glad to see [a] college of New York, at least upon an equality with theirs."

When Samuel returned home to New York in 1765 he pushed officials to open a hospital in his city, one that would outshine the hospital in Philadelphia. Several other physicians joined Bard in his quest, including Peter Middleton, a Scottish physician who had trained in Edinburgh and had performed with John Bard the first dissection in the colonies; Samuel Clossy, an anatomist from Ireland who had published a seminal work on human diseases in 1763; and John Jones, a pioneering surgeon who wrote the first surgical textbook in America. With the weight of such renowned physicians behind the push for a hospital, the Crown in 1771 granted a royal charter to establish the "Society of the New York Hospital in the City of New York in America."

John Jones proved particularly important in the design of the hospital. After graduating from the medical school at the University of Reims, in France, he toured several hospitals in and around Paris, and he hated what he saw. Many European hospitals at the time were squalid, lice-ridden, and grotesquely unhygienic places. "Hospitals were breeding grounds for infection," wrote medical historian Lindsey Fitzharris, "and provided only the most primitive facilities for the sick and dying, many of whom were housed on wards with little ventilation or access to clean water. As a result of the squalor, these places became known as 'Houses of Death.'" Jones found hospital wards in Paris badly designed and utterly demoralizing in execution. He cited a hospital located on Île de la Cité as being particularly egregious, writing:

> In Paris it is supposed that one third of all who die there, die in hospitals. The Hotel-Dieu—a vast building, situated in the middle of that great city—receives about twenty-two thousand persons annually, one-fifth of which number die every year. It is impossible for any man of humanity to walk through the long wards of this crowded hospital without a mixture of horror and commiseration at the sad spectacle of misery which presents itself. The beds are placed in triple rows, with four and six patients in each bed; and I have more than once in the morning rounds, found the dead lying with the living.

Jones wanted to ensure that New York Hospital's wards would be more modern in every respect and that they would encourage healing, not foster infections. "The principal wards," said Jones, "which are to contain no more than eight beds, are thirty-six feet in length, twenty-four feet wide, and eighteen high. They are all well-ventilated, not only from the opposite disposition of the windows, but by proper openings in the side walls, and the doors open into a long passage or gallery, thoroughly ventilated from north to south."

The cornerstone for the new hospital, a small, two-story, H-shaped building, was laid in 1773 on a parcel of land west of Broadway, between today's Worth and Duane Streets. The building was set about thirty yards back from Broadway to provide space on all sides for future growth. The building was nearly finished and ready for patients by the end of February 1775 when it took fire from British troops, setting off a blaze that destroyed all but the exterior walls. Part of the interior had been rebuilt in time for the Continental Army to appropriate the hospital at the start of the Revolutionary War.

Soldiers dug a trench seven feet deep and twelve feet wide around the building as a defense against attack. A surgeon from Rhode Island named Solomon Drowne spent a busy evening on July 13, 1776, applying a "poultice to a man's foot over which a gun carriage ran yesterday" and then performed the first surgery at the makeshift hospital, amputating the arm of a wounded Patriot soldier. Drowne nearly needed his own services when a cannonball from a British ship in the harbor came bounding through the hospital yard and landed just a few feet away. "I believe," he wrote, "it was a 12-pound shot." A month later, the British Army seized the building, and from that point until the end of the war, it was used as a military hospital and barracks for British and Hessian soldiers.

The end of the war brought an opportunity to redefine King's as an American institution. James Duane, New York's Whig mayor and a member of the college's board of trustees, pushed for a name as well as a new charter, a more American charter, the royal charter being "inconsistent with that liberality and that civil and religious freedom which our present happy constitution points out." Duane's vision for the new charter would create a university system with a board of regents at its head. The state legislature eventually passed an act that renamed the school to Columbia College but also gave the state, not the college, complete control.

John B. Pine, a long-time trustee of Columbia University, wrote that the act "robbed the college of its property and franchises and abolished its governing board." Three years later, in 1787, Hamilton, Duane, and Jay proposed to the legislature a measure to restore the rights and privileges granted to the trustees by the charter in 1754, "thus undoing," said Pine, "the wrong perpetrated by the Legislature in 1784 and continuing the corporate existence of the college founded in 1754."

Although Columbia College itself reopened shortly after the war ended, New York Hospital didn't reopen quite as soon, not until 1791. Before that it was used for several years, including 1788, as an anatomy lab, with the subject taught by a surgeon named Richard Bayley. Born sometime in 1745 to William and Susanne Bayley of Fairfield, Connecticut, Bayley had been apprenticed in 1766 to John Charlton, a notable physician in New York City. Three years into his apprenticeship, he married Charlton's sister, Catherine. The pair would go on to have three children, one of whom, Elizabeth Ann (Bayley) Seton, would become a nun, establish the Sisters of Charity of St. Joseph, and, in 1976, become the first American to be canonized as a saint.

Early in the 1770s, not long after Bayley married, he traveled to London, where he met the great William Hunter, who by that time had become a world-famous obstetrician and surgeon. Hunter granted the young Bayley access to his new anatomy lab at the corner of Archer and Great Windmill Streets in central London. Bayley spent a year with the esteemed Hunter before returning to New York, where he opened a practice with his brother-in-law and began focusing on the causes and treatment of croup, a childhood respiratory disease little understood at the time. Bayley traveled to London several more times to study with Hunter. On his last trip, he set sail on one of Admiral Richard Lord Howe's warships, where he served as an onboard surgeon. The ship landed at Staten Island on July 12, 1776, just eight days after the start of the American War of Independence.

Bayley was stationed with British troops in Rhode Island when he learned that his wife had become seriously ill. The only way he could visit her was to resign, which he did in 1777. He returned to New York just in time to see his beloved Catherine pass away. He then began a surgical practice and taught anatomy to interested members of the public.

When the war ended, Bayley became caught up in anti-Loyalist sentiments raging through the city and faced accusations of having conducted experimental surgery on prisoners when he was stationed in Rhode Island. So incensed were the accusers that they demanded he be banished from the country. New York's chief justice, William Smith, investigated the accusations and, at some point, asked a Patriot physician he knew, Silas Holmes, to describe his dealings with Bayley in Rhode Island in 1776. "When I first saw Doctor Bayley," said Holmes, "he expressed sorrow that the wounded had been so long neglected. He declared that nothing in his power should be wanting to make them as comfortable as their situation would admit of." Holmes went on to explain that the overall conduct of Bayley "was not only such as we had a right to expect from a generous enemy, but such as would meet with approbation from a friend." Smith thus deemed the accusations against Bayley unfounded. His name now cleared, Bayley began lecturing on surgery in 1785 and, with his physician son-in-law Wright Post, teaching anatomy in a section of New York Hospital.

To educate students properly, to show them the inner workings of the human body and the many unique intricacies of human anatomy, Bayley and Post needed bodies. Here is where Bayley's training in London with William Hunter came to the fore, not just in dissection but also in dealing with body snatchers and in the value of collecting bones, skulls, internal organs, and other body parts along the way.

Somewhat shorter in height than most, slim, with a straight nose, strong jaw, and genial brown eyes, Hunter lived simply and plainly. He enjoyed visiting coffeehouses and sipping a glass of claret at pubs. His brother, John, who became William's assistant and an arguably even better anatomist, once said of his brother's penchant for saving that he was a "miser." Another physician, one Dr. Manning, described Hunter as "an ill-natured waspish creature."

Hunter was a gifted lecturer and was roundly praised as an entertaining speaker who could recall fascinating stories on the fly to help illustrate a point. Biographer Sir Henry Halford said of Hunter, "In his lectures, he was celebrated for the variety and appositeness of the anecdotes with which he enlivened or illustrated the theme. Men of the world, artists, and persons of every taste listened to him with gratification."

At one point, Hunter decided to retire from lecturing, but his students protested enough that he relented. "A man may do infinitely more good to the public by teaching his art," Hunter said, "than by practicing it." Matthew Baillie, Hunter's nephew and renowned anatomist in his own right, summarized his famous uncle this way:

No one ever possessed more enthusiasm for his art, more persevering industry, more acuteness of investigation, more perspicuity of expression, or indeed a greater share of natural eloquence. He excelled very much any lecturer whom I have ever heard in the clearness of his arrangement, the aptness of his illustrations, and the elegance of his diction. He was perhaps the best teacher of anatomy that ever lived.

Hunter drew on his extensive collection of body parts to help students focus on specific parts during his lectures. He had been collecting body parts for decades, many of which came from bodies stolen from local cemeteries and dissected in his anatomy lab.

His collection had grown so large that in 1770 he moved from his home at 49 Jermyn Street, near Piccadilly Circus, to a much larger and grander home on Great Windmill Street. In it was "a handsome amphitheater and other convenient apartments for his lectures and dissections," as well as "one magnificent room … fitted up with great elegance and propriety as a museum."

Hunter understood the moral and ethical morass created by exhuming bodies for his students, and he knew that body snatching had been a criminal offense in the United Kingdom since 1752, but his students needed bodies. He obtained bodies legally here and there, whenever possible from patients who had died from childbirth or criminals who had been executed. Dissecting the body of an executed prisoner was allowed under the law and had been for many years. The legal supply, however, never met the demand. So, like others before him, Hunter turned to resurrectionists to fill the gap.

He tried to remain discreet in his dealings with the pillagers, but eventually, the nighttime goings on at his stately home attracted attention. An enterprising journalist from a local newspaper wrote about the unusual number of bodies being dropped off at 16 Great Windmill Street and, even worse, that some of the remains were being thrown down a well in the garden. The stories sparked such fury that groups of enraged citizens occasionally rioted outside Hunter's house. Cartoonists joined in, with one depicting Hunter "hastily abandoning a woman's body on sighting an approaching night watchman."

Snatching newly buried bodies had been a regular medical school practice throughout Europe for two centuries. Joseph Guichard Du Verney, a pioneering physician of the seventeenth century, used resurrectionists frequently. Du Verney studied diseases and disorders of the ear and wrote a detailed, illustrated text on the ear that physicians then and now consider "remarkable not only for its anatomical presentations, but also for its author's thoughts on the physiology and pathology of the ear." He was able to present such detailed knowledge partly, or perhaps mostly, because he dissected

Caricature of William Hunter and a body snatcher being run off by a night watch-man. COURTESY OF CARL H. PFORZHEIMER COLLECTION OF SHELLEY AND HIS CIRCLE, *THE NEW YORK PUBLIC LIBRARY.* "THE ANATOMIST OVERTAKEN BY THE WATCH IN CARRYING OFF MISS W—TS IN A HAMPER," NEW YORK PUBLIC LIBRARY DIGITAL COLLECTIONS.

bodies taken from the Clamart cemetery outside of Paris, "much to the alarm of the population, which viewed such spectacles with horror."

Noted Swiss anatomist Albrecht von Haller was a student in Paris when he found himself paying ten francs to purchase a body from a gravedigger. A Dublin cemetery called Bully's Acre was routinely pilfered of 1,500–2,000 bodies a year in the late 1600s, and one was the body of English author Laurence Sterne. A close relative of Sterne's was in a dissecting room shortly after the writer's death in 1678 and suddenly recognized his kinsman on the table.

Richard Bayley's time with Hunter must have given him great insight into resurrectionists and the benefits of possessing his own collection of specimens. Exactly when Bayley started collecting body parts in New York is unknown, but by 1788, he had amassed an impressive number of specimens and stored them in his anatomy lab at New York Hospital, the lab that overlooked a large lawn where children often played.

As Bayley was studying with Hunter in London, a young Alexander Hamilton was preparing to start his first year at King's College. For at least part of that year, 1773, Hamilton thought he wanted to become a doctor, so he took anatomy classes taught by Clossy, one of the men behind the campaign to build a city hospital. "I have often heard [Hamilton] speak of the interest and ardour he felt when prosecuting the study of anatomy," recalled Hamilton's physician, David Hosack. "Few men knew more of the structure of the human frame and its functions [than Hamilton]."

Hamilton soon decided that a medical career was not for him, possibly because his energies were being directed elsewhere. He had entered King's around the time of the Boston Tea Party, a demonstration against the draconian measures of the Stamp Act. The next several months proved particularly exciting for the young man. "These were stirring days for Hamilton," said Chernow, "who must have been constantly distracted from his studies by rallies, petitions, broadsides, and handbills." Hamilton decided to speak at a rally held July 6, 1774, on a grassy area near King's called the Commons, now City Hall Park. Perhaps Hamilton had already made up his mind to discontinue his anatomy studies by that time, but if not, the speech he made that day surely removed whatever doubts he had. He stood before the assembled and spoke extemporaneously, starting off timidly but then building in intensity, showing hints of the oratorical skills he would soon hone to perfection.

Hamilton supported a boycott of British goods, urging the colonies to "stop all importation from, and exportation to, Great Britain" until the Crown reopened Boston Harbor. He argued against exorbitant taxation and called for the colonies to come together in this difficult time. Doing so, he said, "will prove the salvation of North America and her liberties." He noted that without unity, "fraud, power, and the most odious oppression will rise triumphant over right, justice, social happiness, and freedom." His oration at the Commons became known as the "Speech in the Fields" and marked the moment he became one of the prominent voices of the revolution.

A colleague of Hamilton also possessed an outsized voice, though not in the same way. Hamilton had been serving as Washington's aide-de-camp when a forty-seven-year-old Prussian officer arrived at Valley Forge, Pennsylvania, in the glacial winter of 1777. He was a portly fellow with a conspicuous nose, rounded chin, and dark eyes that could penetrate an insincere mien. Born into a military family, the man had been raised in the midst of "guns, drums, trumpets, fortifications, drills, and parades," comfortable in the art and science of war. He and his aides had landed in Portsmouth, New Hampshire and then traveled to York, Pennsylvania, where Congress had been meeting. He gave the members a letter of introduction from Benjamin Franklin, which announced him as "His Excellency, Lieutenant General von Steuben, Apostle of Frederick the Great." In reality, he was a captain, not a general.

Baron Friedrich Wilhelm von Steuben had become an officer in the Prussian army at age seventeen and had fought against Austria and Russia during the Seven Years' War, a worldwide conflict known in the United States as the French and Indian War. He lost his commission in 1763, having been a casualty of a bitter rival and a peacetime reduction in troop strength. He found a position with a tiny German principality, but his salary was slashed, so he began searching for an army to join. He applied for a position in the German army at Baden but was rebuffed when rumors swirled that he had "taken liberties with young boys." No evidence to support that rumor has ever surfaced, but that didn't stop it from destroying Steuben's reputation in Europe. "Rather than stay and provide a defense," said historian William E. Benemann, "rather than call upon his friends . . . to vouch for his reputation, von Steuben chose to flee his homeland."

Friedrich Wilhelm von Steuben in a painting by Ralph Earl. GIFT OF MRS. PAUL MOORE IN MEMORY OF HER NEPHEW HOWARD MELVILLE HANNA JR., B.S. 1931.

Franklin and four French leaders—Silas Deane, Charles Gravier (comte de Vergennes), Claude-Louis (comte de St. Germain), and Pierre-Augustin Caron de Beaumarchais—wrote glowing letters supporting Steuben's enlistment. With that support, Congress immediately assigned him to Washington's Continental Army, and off he went, riding an imposing horse with a Russian wolfhound by his side. He wore a grand uniform with a Star of the Order of Fidelity on his left chest. The medal, awarded to him in 1769 by the niece of King Frederick the Great, was a prestigious medal in Germany and featured a Maltese cross with a gold finial on each of its eight points. The white enamel center showed the word "FIDELITAS" at the top and a trio of mountains below.

He must have made an imposing sight along the roads from York to the camp at Valley Forge, reminding one soldier of Mars, the god of war. "The trappings of his horse," recalled the soldier, "the enormous holsters of his pistols, his large size, and his strikingly martial aspect all seemed to favor the idea." Washington was also impressed, saying that Steuben "appears to be much of a gentleman, and as far as I have had an opportunity of judging, a man of military knowledge and acquainted with the world."

Steuben had been accustomed to soldiers following his orders, but in Valley Forge, he faced troops who didn't speak German. Steuben spoke French but little English, so his military secretary, Pierre-Etienne Du Ponceau, translated his orders from French to English, with help from Hamilton and John Laurens, Washington's aides-de-camp. Steuben's written orders were clear and professional, rather docile compared with verbal orders given when angry. Steuben had taken over training twenty-odd battalions of weary troops, hungry troops wearing ragged uniforms, many of them without boots to shield their feet from the biting cold. No wonder he faced times of frustration, despair, and sometimes anger over his troops' seeming inability to follow his commands. Sometimes he ended up cursing at them. He didn't know any curse words in English, though, so he would turn to one aide or another. "My dear Du Ponceau," he would say, "come and swear for me in English. These fellows won't do what I bid them!"

Steuben taught his new troops how to march, use a bayonet properly, communicate commands on the battlefield, and fire their weapons more rapidly than they ever could before, essential for the kind of direct combat they faced against British troops. He greatly improved sanitation in the camp, placing kitchens on the opposite side of the camp as the latrines, which he ordered placed on downslopes. His organizational skills proved exemplary, as did his ability to simplify directives and engender respect and support. He also wrote *Regulations for the Order and Discipline of the Troops of the United States*, commonly called the "Blue Book" for its blue cover. The book standardized training methods for the entire army and became arguably the most important military factor in the war. The US Army continued using the book for training until 1814.

By the time the war ended, Steuben had become one of the nation's preeminent figures, second, perhaps, only to Washington himself. He resigned his commission early in 1784 and began renting a house on 57th Street in New York City the following October. He was visited by a few aides-de-camp he had worked and lived with during the war, including James Fairlie, William North, and Benjamin Walker. Biographer Joseph Doyle said that "Steuben was not able to entertain his friends in his own house very long, probably not over three years, and perhaps less, for by that time the pittance he had received from the Government was exhausted."

From there he moved in with friends, first with Benjamin Walker and his wife in their house on Maiden Lane, and then with a Dr. Vache on Courtlandt Street, where he "took his meals with Misses Dabeny, who kept a popular boarding house in Wall Street." He lived there throughout the spring of 1788, in the Courtlandt Street house, just a few blocks south of Columbia College, and so was close by when he heard in April about a mob attacking New York Hospital.

CHAPTER 4
DISSECTION THROUGH THE AGES

LEONARDO DA VINCI WAS FIFTY-SIX YEARS-OLD AND A RENOWNED artist when, in 1508, he sat at the bedside of a frail, elderly gentleman at the Hospital of Santa Maria Nuova in Florence. "And this old man," Leonardo wrote, "a few hours before his death, told me that he was over a hundred years old, and that he felt nothing wrong with his body other than weakness. And thus, while sitting on a bed . . . , without any movement or sign of any mishap, he passed from this life."

That night, or perhaps the next day, Leonardo anatomized the man's body in the hospital's crypt, poised to learn, in his words, "the cause of so sweet a death." Although Leonardo had dissected individual limbs before, as well as the bodies of birds, oxen, and horses, the centenarian became the first full-body human dissection Leonardo had ever performed. After the autopsy he deduced that the man had died from "a fainting away through lack of blood to the artery which nourishes the heart and the other parts below, which I found very dry, thin and withered." Leonardo also observed that the man had suffered from cirrhosis of the liver, describing the liver tissue as "desiccated and like congealed bran, both in colour and substance," and that when touched the tissue "falls away in tiny flakes like sawdust, and leaves behind the veins and arteries."

Leonardo performed as many as thirty dissections in his life, each time making precise illustrations of what he saw. "Mustering all of his draftsman's techniques," wrote biographer Walter Isaacson, "he made detailed underdrawings in black chalk, then finished them with different colors of ink and washes. With his left-handed curved

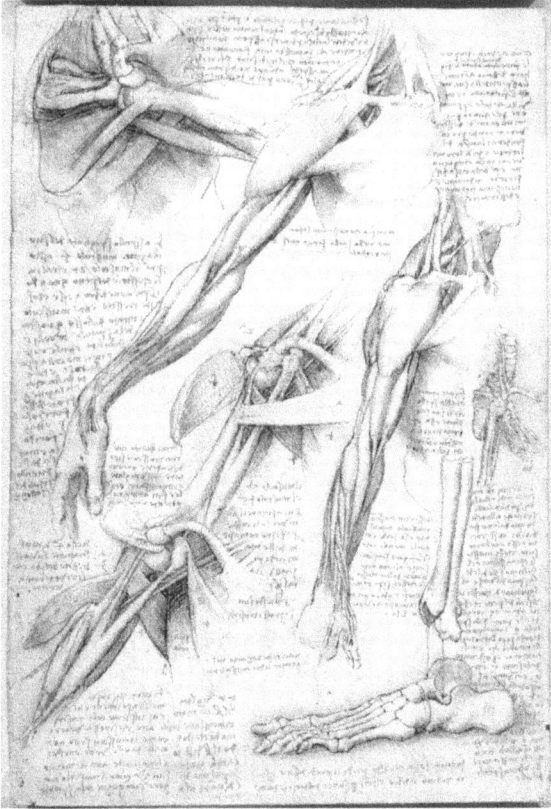

Notes and drawings of the muscles of the arm and shoulder and the bones of the foot by Leonardo da Vinci, who based the images on a dissection he performed on the body of a one hundred-year-old man, circa 1508. Leonardo possessed a remarkable ability to remember details and could draw nuanced illustrations from memory. COURTESY OF THE ROYAL COLLECTION TRUST.

hatching lines, he gave shape and volume to the form of bones and muscles and with light lines added the tendons and fibers. Each bone and muscle was shown from three or four angles, sometimes in layers or in an exploded view, as if it were a piece of machinery he was deconstructing and delineating. The results are triumphs of both science and art."

Leonardo lived during a time when autopsies were performed for the same reasons they're performed today: to study human anatomy and to find out how and why a person died. Postmortem dissection was allowed, and at times even encouraged, from the thirteenth to the sixteenth centuries in Europe, with dissections recorded as early as 1240 CE in Italy and the 1500s in France, Germany, and other countries. Most dissections happened during the winter, when colder weather helped to preserve bodies longer. Carnival season during the latter half of January was particularly popular, one spectacle drawing attention away from another: dissections. Dissections tended to attract not just students studying to become doctors but also regular citizens. Subjects came mostly from prisons, and it helped if bodies were of tall, thin men; organs in their body cavities would be more visible to whoever was watching than would organs in an obese person.

Studying anatomy through dissection was considered at the time a moral duty. "Anatomy typically had the blessings of religious authorities," said author Michael Sappol. "In some circles, dissection was even regarded as a pietistic activity: in studying the human body, we see God's handiwork." Not until noted anatomist and physician Andreas Vesalius came into the picture did dissections take on a more desperate aspect. "He and his students forged keys," said historian Katharine Park, "rifled tombs and gibbets [cages to display executed criminals], and stole in and out of ossuaries in a series of nighttime escapades that he recounts with evident relish and amusement, particularly when female bodies were involved." Vesalius might even have autopsied still-living beings. He wrote once, in describing apparent moisture in the membranes of the heart, that he was "eager" to actually view the moisture. Park wrote that "he opened the body of a man who had just died in an accident and took out what he described as 'the still pulsing heart.'"

Vesalius spurred a paradigm shift in the way medical students learned anatomy. He had studied medicine in Paris and then moved to Padua to teach anatomy. He disliked the traditional manner of conducting a dissection. In those cases, an anatomy professor, or *lector*, would read from an anatomy textbook—commonly *Anathomia*, by Medieval anatomist Mondino de Liuzzi, a fundamental medical text at the time—while someone else, the *ostensor*, would stand beside the body and point out which body part should be dissected. The *sector*, normally a barber surgeon, would then dissect the body part. Vesalius decried this "hateful method," saying that "One dissects the body and another describes its parts: the first, perched on a pulpit like a crow, haughtily repeating ideas that he didn't learn directly from the cadaver, but that he read in other's books." Vesalius went on to publish, in 1543, *De Humani Corporis Fabrica Libri Septem*, a textbook that would stand as the epitome of anatomy texts for more than three hundred years.

Dissections were also carried out for legal reasons, such as determining whether someone had been murdered or died of natural causes. During the sixteenth century, the law required that any pope who died, regardless of circumstance, must be dissected to ensure they weren't poisoned while in office. Pope Leo X was one of them. Leo X reigned from 1513 until his unexpected death in 1521. The pope's assistant, Paride de Grassi, described the holy dissection: "The pope's body was opened, and his heart and spleen were found to be corroded and the spleen similarly partially cauterized, which the surgeons and physicians viewed with amazement." The pope, they decided, had been poisoned.

Determining sainthood could also involve dissection, as was the case with Sister Chiara of Montefalco, abbess of a monastery in central Italy, who died August 17, 1308. Chiara's fellow nuns decided that her body "should be preserved, on account of her holiness and because God took such pleasure in her body and her heart." Sister Francesca performed the dissection, with Sisters Caterina, Lucia, and Margarita assisting. Francesca "cut open the heart with her own hand, and opening it they found in the heart a cross, or the image of the crucified Christ." Kalina Yamboliev, history professor at the University of California, Santa Barbara, explained that "the heart was considered to be the place where God inhabited the body, while

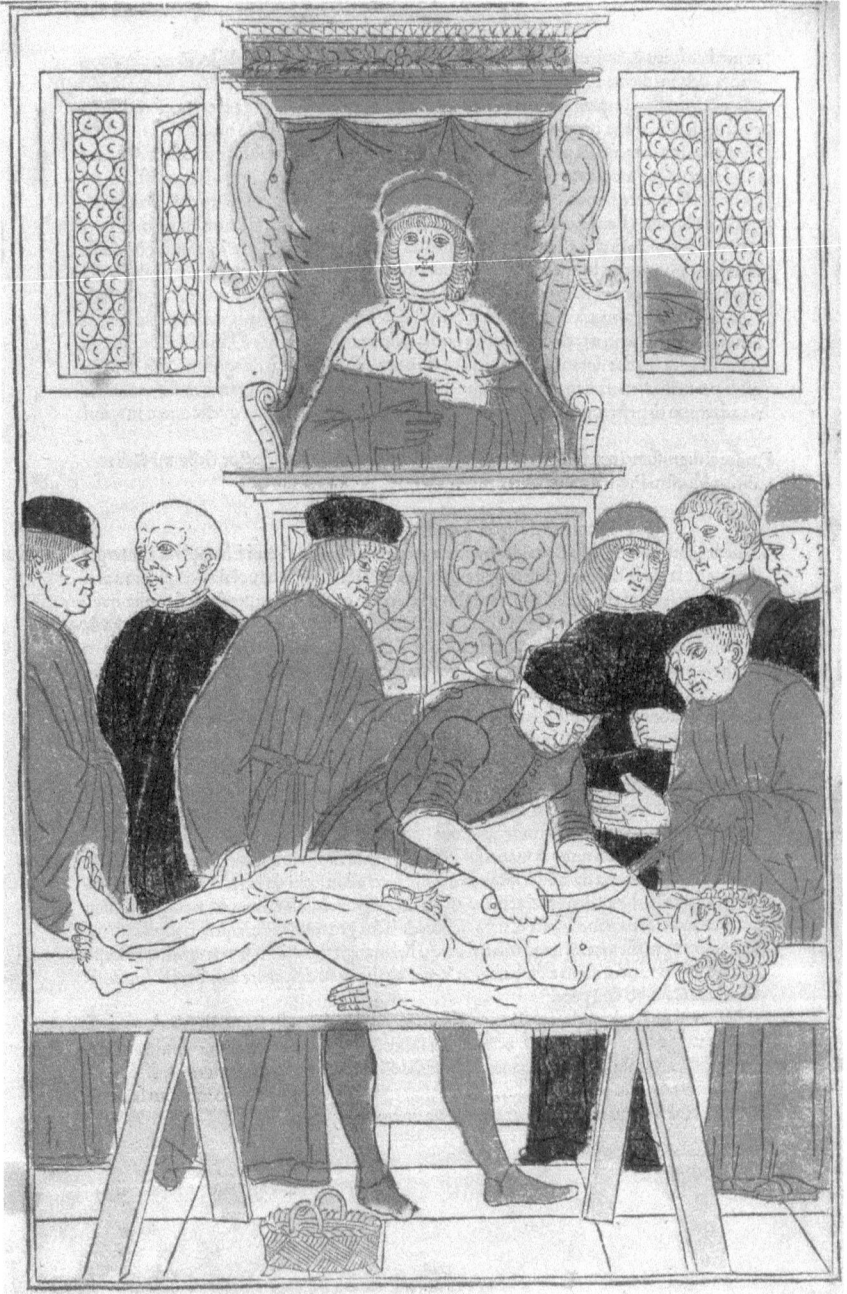

A woodcut from a volume of a medieval medical book, *Fasciculus medicinae*, showing a dissection. A lector (standing, with hat) watches over a barber surgeon working under the direction of an ostensor (right, pointing to cadaver).

demons were thought to reside in the bowels. Women were considered particularly prone to possession by these sacred forces."

Sister Francesca also found four structures in Chiara's heart that reminded her of a crown of thorns, a whip, and three nails, items considered holy from the time of Jesus's crucifixion. The nuns took the appearance of those structures, together with three stones discovered in the nun's gallbladder, which they believed were a sign of the Holy Trinity, as signs of God's presence. Many years later, during the canonization process, Francesca's testimony proved critical in making Sister Chiara a saint.

Saint Chiara was one of several saints whose autopsy results had proved essential in the canonization process. Blessed Margerita of Citta' di Castello had lived a difficult life. Born blind and with physical deformities, Margerita was hidden from public view as a child and eventually abandoned by her parents. Pious and devoted to others, she died at age thirty-three in 1320. Viewers examining her heart found three small stones, "each impressed with unique images," said Yamboliev, "including depictions of Mary, the baby Jesus, Joseph, the Holy Spirit, and a repentant who was considered to be Margerita herself. A modern thinker will once again search for more practical explanations—scars and marks on her heart, or clumps of tissue that would appear like small stones—but to Margerita's contemporaries, these objects were clearly marks of the divine."

Pope Clement VII officially decreed in 1537 that the church would accept dissection for anatomy education. Following that decree the number of dissections increased sharply, as did the public's infatuation with viewing them. As a result temporary anatomy theaters, called *theatra anatomicum temporaria*, soon popped up to showcase dissections. Former director of anatomy at the University of Rostock, Gert-Horst Schumacher, described the basic style of one of those theaters:

It should be a sizeable and well-ventilated place with seats all around it, and of such a size as to hold a great number of spectators, so that the

dissectors shall not be disturbed by the crowd. Seating should be allot-
ted in order of rank. There must be an usher to keep an eye on every-
thing and to put people in their places, as well as guards to restrain the
eager public as it enters. Two reliable stewards should be chosen to make
the necessary payments from the money that was taken. Torches were
ordered, because the body must be in sufficient light.

By the end of that century, permanent theaters had been built specifically to display dissections. Padua and Bologna, Italy, each had a *theatrum anatomicum.* Theaters were also built in Paris, Berlin, Uppsala in Sweden, and Amsterdam, Copenhagen, and Leiden in Netherlands. The Leiden theater, opened in 1594, was a "circular amphitheatre with six tiers around a rotatable dissection table, adorned with human and animal skeletons, and accommodated in the apsis of a secularised church." Public dissections at this theatre, said Tim Huisman, curator at the National Museum Boerhaave, "were conducted with great solemnity and decorum, almost like religious ceremonies. They were attended by the burgomasters of Leiden and by the senate of the University, and all lectures and other academic activities were suspended when these anatomical demonstrations were held."

Interior of the Anatomical Theater in Archiginnasio, a palace in the heart of Bologna's historic city center. COURTESY OF THE MUNICIPAL LIBRARY OF BOLOGNA.

The theater included a rotating dissection table, the better to display the body to the assembled students and the growing number of citizens who longed to view such scenes. Dissections were in many ways performances conducted by a physician who sat at the head of the table and instructed an anatomist to dissect this body part or that.

Attending a dissection in the Padua theater, the oldest surviving anatomical theater in Europe, meant standing hip-to-hip along narrow rows built almost one atop the other. Each successive row of viewers would have to lean over people in the row below to see the dissection table. Padua's theater could hold about two hundred people in six concentric rows, or circuits. The first circuit was typically reserved for medical college faculty, senior members of the surgeons' guild, local judges, and nobles in the area. Next came apothecaries, physicians, and surgeons; then medical students, usually in the third and fourth circuits. Members of the paying-public took whatever space was left over.

In Leiden guests were ushered to their seats by a *praefectus*, a special assistant to the teacher, while other assistants, the *custodes*, served as barriers to entry for the ticketless public. Tickets for seats near the table, aside from those reserved for faculty and other esteemed guests, cost more than tickets at the top of the amphitheater, the cheap seats. Music played in the background. Candles held by students often provided the only light for the anatomists. Because dissections occurred over several consecutive days, perfumes sprayed in the hall helped to mask odors of decaying flesh.

During the eighteenth century in Europe, the number of students wanting to become physicians increased substantially, and they all needed training in anatomy. In Scotland, where barely any dissections were carried out between 1500 and 1670, dissection was allowed on just one executed criminal per year. Then, in 1674, the Edinburgh town council, recognizing the need to obtain more bodies for anatomy study, granted permission to its surgeons to dissect not just the bodies of executed criminals but also the bodies of

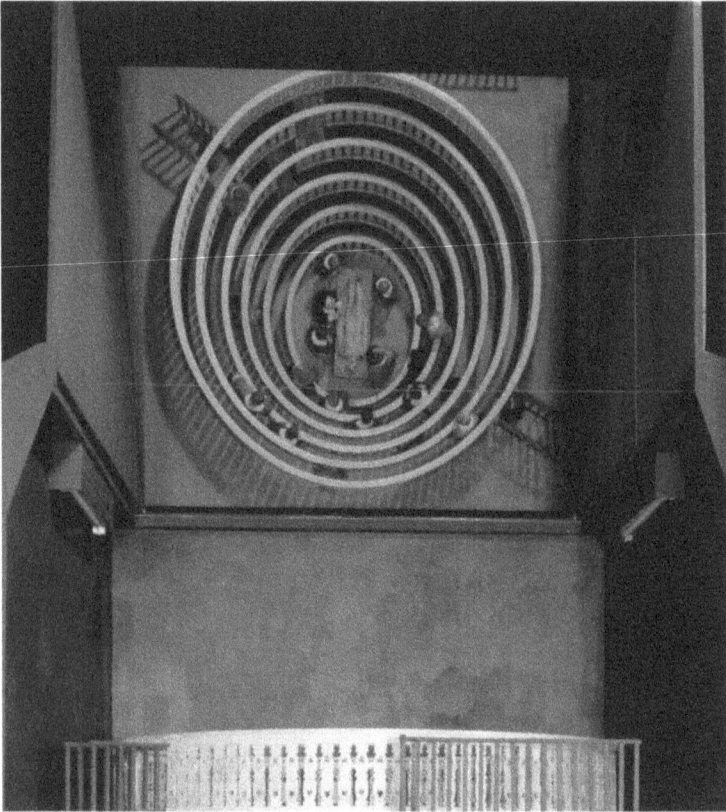

View from above of a scale model of the anatomy theater at the University of Padua. William Harvey, who discovered how blood circulated in the human body, attended the university and watched dissections here.
COURTESY OF THE SCIENCE MUSEUM GROUP COLLECTION.

orphans, stillborn babies, people who had committed suicide, and those who had died in prison.

That permission, however, came with two stipulations: that Edinburgh must build an anatomical theater within three years and that one dissection for the public must be held each year. Construction of the theater finished in 1697, but the first dissection wasn't performed until 1702. The body in that dissection was anatomized in eight sections over eight days, each section dissected by a different professor.

From that point on the number of students flocking to Edin-burgh exploded. Many of them wanted to study with one of the three famous Monro anatomy professors—Alexander *primus*, who founded the medical school at the University of Edinburgh and became highly respected throughout Europe for his expertise; Alex-ander *secundes*, who continued his father's preeminence in anatomy education and expanded it further with several noteworthy treatises on common diseases; and Alexander *tertius*, who became an anatomy professor at the university following his father's retirement. Partly as a result of the fame of the Monro professors, Britain became the center of medical studies in Europe, particularly Edinburgh and London, each of which possessed outstanding medical schools.

Arthur M. Lassek, a neurologist and author of *Human Dissec-tion: Its Drama and Struggle*, explained that physicians and surgeons began to realize that anatomical knowledge was "founded on inse-cure grounds, that it was based on equivocal tradition rather than on scientific considerations, and that ignorance of specific facts was retarding the whole of medicine." Surgeons recognized that they needed a far greater understanding of human anatomy if medicine and surgery were ever to advance. Universities began hiring more anatomy and surgery professors, and experienced surgeons began opening their own clinics and schools. Several private professors became instrumental in advancing the teaching of anatomy, includ-ing Percivall Pott, the first professor of anatomy at London's Com-pany of Surgeons, and William Hunter, whose private school on Great Windmill Street offered innovative approaches to anatomy instruction.

To truly advance medicine, though, bodies would be needed, more bodies than the laws allowed. The British government and its people seemed torn. Much of the public hated the idea of dissection. They believed that dissections would leave their loved ones' body mangled and that their soul might not be raised to heaven on judg-ment day. Many also considered the dissecting act itself disgraceful, a naked body lying on a slab for all to see. Yet some people could understand the need for surgeons to better understand the human body and that the only way to do that was to examine cadavers.

London chose to clamp down on dissections and pass the Murder Act of 1752, officially called the Act for Better Preventing the Horrid Crime of Murder. The act authorized members of the Company of Barber Surgeons and the Royal College of Physicians—and only those members—to perform autopsies, and even then dissections could occur only on "a very limited number of human cadavers," mostly executed criminals. The act decreed that the body of any criminal executed "in the county of Middlesex, or within the city of London or the liberties thereof" would be "immediately" transported by sheriffs to Barber-Surgeons' Hall, where the body would be "dissected and anatomized by the said surgeons, or such person as they shall appoint for that purpose." So many bodies of executed criminals were dissected throughout Britain, said one historian and anatomy professor, that "anatomical dissection became synonymous with capital punishment."

Judges could choose to have an offender's body hung in chains after execution if the crime was deemed "most atrocious," but otherwise, according to the law, the body must be dissected. Under no condition could the body be buried until after the surgeons were finished with it. The act further stipulated that anyone attempting to "rescue" the body of an executed criminal from an anatomist could be "transported to some of his Majesty's colonies or plantations in America for the term of seven years." How many rescuers were banished to the colonies is unknown, but it seems that being sent across the Atlantic was considered a satisfying enough punishment.

Human dissection occurred in the colonies before the mid-1700s, but scant evidence exists about it. According to Lassek, "Records of such events are rare because there were no medical colleges or journals which existed during the period. The only other source of communication was by means of a few newspapers, which largely excluded such items, because of public antipathy to the procedure. There was probably more anatomizing going on than can be revealed."

Historians differ about when the first dissection in the colonies was performed and who performed it. Francis R. Packard, who put together the enormous *History of Medicine in the United States*, claims that the first dissection occurred in 1674. Packard said that a man named John Josselyn, in "An Account of Two Voyages to New England," published in 1674, wrote:

A young maid that was troubled with a sore pricking at her heart, still as she leaned her body or stept down with her foot to the one side or the other; this maid during her distemper voided worms of the length of a finger, all hairy with black heads; it so fell out that the maid dyed; her friends desirous to discover the cause of the distemper of her heart, had her opened, and found two crooked bones growing upon the top of the heart, which as she bowed her body to the right or left side would jab their points into one and the same place, till they had worn a hole quite through.

Lassek believes, however, that the first dissection occurred even earlier, when Giles Firmin, a Boston physician and minister, anatomized a man in 1638, probably an executed prisoner. Regardless, according to early legal codes, such as a 1641 Massachusetts code known as "Body of Liberties," the body of any executed criminal could be dissected legally. Dissections were probably performed by doctors, who at that time were either self-taught or apprenticed to a surgeon or barber-surgeon from England, of which there were precious few in the land across the pond. The first European medical man to step ashore in the New World was Dr. Thomas Wotton, surgeon-general of the colony, who came to Jamestown in 1607 aboard the first ships from England.

A Londoner named Daniel Turner would be granted the first medical degree in the colonies. Turner had completed a seven-year medical apprenticeship in London and was admitted in 1700 to the Company of Barbers and Surgeons. He wanted to be accepted as a member of the Royal Society of Medicine, but he "lacked the personal connections, social status, university education, or recognition as a scholar" for admission. Seeking to elevate his standing in the London medical world, he sent a letter to Yale College, along with a set of twenty-five books from his own library. "If Your Lordships judge me worthy of the Degree of Doctor of the Yale Academy," read the letter, "and care to transmit to me a diploma, I shall accept it not only as a token of your gratitude, but shall consider it an honor as great as if it had been conferred by another, even more renowned university." For reasons still unknown, the college awarded an honorary MD degree to Turner in 1723, a barber-surgeon who hadn't attended Yale or even set foot in the colonies.

Barbers and surgeons in Britain had a long and complicated relationship until the mid-1700s. Barbers, or barber-surgeons, were the medical practitioners of the Middle Ages. They were of necessity skilled at using sharp instruments and became adept at not just cutting hair and beards but also performing tooth extractions, bloodletting, leeching, and amputations. Two different groups evolved, the Company of Barbers and the Fellowship of Surgeons, which merged in 1540 to form the Company of Barber Surgeons. Two centuries later, the group split up, with surgeons forming what eventually became known as the Royal College of Surgeons.

Surgery was then and for a long time after considered a lesser calling than that of medicine, especially in Britain, less so in the colonies. "In the England of 1700," said medical historian Richard H. Shryock, "the London College of Physicians was authorized to control licensing. This elite body limited its certification to the graduates of Oxford and Cambridge, and so never approved enough men to meet the needs of a tenth of the population. The consequent vacuum was partly filled by licentiates of the apothecaries guild, and by the 1700s apothecaries made up the ranks of ordinary practitioners. Surgeons, overseen by the Surgeon's Guild, were viewed as an inferior group in comparison with the licensed physicians." Even today in England people address male surgeons as *Mr.*, leaving *Dr.* for non-surgeon physicians. Another medical historian, Irvine Loudon, said that "surgeons, or rather male surgeons, are always addressed as *Mr.* in the United Kingdom and the Republic of Ireland, sometimes but not always in Australia and New Zealand, and rarely in Canada or the United States." The appellation *Mr.* is today a badge of honor for British surgeons.

A similar dichotomy arose in America as well, the same concepts having been brought to the colonies by European physicians. Said Shryock, "Surgeons, overseen by [England's] Surgeon's Guild, were viewed as an inferior group in comparison with the licensed physicians. Since there was no real interference with all sorts of irregulars and quacks, these various forms of licensing meant little in practice. Hence it is not strange that, in the distant colonies, governmental control over medical practice almost disappeared."

Colonial medicine began to thrive, primarily in three cities: Boston, New York, and Philadelphia. As the 1700s wore on, Philadelphia pushed to the forefront of those cities and became a leading center of medicine and medical education. Leading that charge were William Shippen Jr. and John Morgan, men who started their careers as colleagues and ended up enemies for life.

CHAPTER 5
MEDICAL MELEE

WILLIAM SHIPPEN JR. WAS BORN IN PHILADELPHIA TO WILLIAM Shippen and his wife, Susanna, on October 21, 1736. The elder Shippen, a respected physician in the city, sent his son to West Nottingham Academy for a formal education and religious training. The academy had been started in 1744 by Reverend Samuel Finley, who would one day become president of the College of New Jersey. Shippen graduated from the college in 1754 and gave the valedictory address for his class. He went on to apprentice with his father for four years, after which his uncle funded a trip to England to continue his medical studies. Shippen biographer Betsy Copping Corner described Shippen as "curious, eager, a thirst for knowledge, fond of excitement, fond of people, [and] socially at ease."

During his time in London, Shippen decided to focus his education on obstetrics and to learn anatomy at the hand of John and William Hunter. Shippen applied himself eagerly to his studies, attending lectures, going on doctors' rounds at area hospitals, and conducting dissections from July through December 1759. He decided in early 1760 to finish his studies at the University of Edinburgh and graduated from there in 1761. John Morgan entered the university that same year.

Morgan was born in Philadelphia a year earlier than Shippen, on October 16, 1735. Morgan's grandfather had traveled from Wales with his son, Evan Morgan, to Pennsylvania in 1717 and, at some point, returned to his homeland. He left his family an inscription on their family bible: "I, David Morgan, gentleman of Wales, bequeath to my descendants in America the comfortable certainty: They came neither from kings or nobles but from a long line of true gentlemen and women with unstained names." Morgan attended West Nottingham Academy, as had Shippen, and graduated from the College of Philadelphia in 1757 with a Bachelor of Arts degree. Throughout nearly all of those educational years, Morgan also apprenticed with John Redman, a highly respected physician in Philadelphia. Redman treated Morgan like a son and gave him increasingly more complex procedures to handle, including bandaging wounds, bleeding patients, pulling teeth, and assisting in surgery.

Morgan also spent a year as a regimental surgeon at Fort William Henry, along the banks of Lake George in New York, during the French and Indian War before returning to Philadelphia for graduation exercises on May 17, 1767. From there he sailed to London to further his medical education. He brought with him, tucked in his suitcase, twelve letters for delivery to William Shippen Jr., already ensconced in medical life there, and two letters of introduction, one of which was for Benjamin Franklin, serving as Pennsylvania's agent in Britain. There he studied with William Hunter and John Fothergill, an internationally recognized physician and botanist, who would one day send home to Philadelphia seven cases of illustrations by the greatest artist of human anatomy at the time, Jan van Rymsdyk. Morgan was in London in the fall of 1761 and watched coronation festivities of King George III with Shippen. "Dr. Shippen (who arriv'd from Scotland but the day before) & myself with a couple of ladies," wrote Morgan in a letter home, "were happy eno' to have a good sight of the procession, just opposite to Westminster Hall, which it came out of. I don't remember on any occasion to have ever seen one fourth of the number of people what had crouded [sic] together that day to get a sight of it."

Both men spent more than a month in London that fall. "Morgan inquired eagerly about Edinburgh," said Morgan's biographer Whitfield J. Bell Jr., "the professors, the citizens he should meet, where he might find comfortable lodging in a city notoriously ill provided with accommodations for travellers [sic]. The young men discussed their work—Morgan's in London, Shippen's in London and Edinburgh." They talked about the wealth of opportunities for medical education in Britain and the "pitiful opportunities" in America.

The men discussed how they wanted to improve opportunities for medical training in their homeland and thought that they should found their own medical school. They finally brought their concerns to a mentor they both knew, John Fothergill. The older physician helped the men adjust their soaring goals to more practical ones, and in the end, he cautiously approved of their plans. As Morgan prepared to travel to Edinburgh to complete his training, according to Bell, he added another goal: "to fit himself for a career as a teacher of medicine in his own country."

Both men returned to America competent, experienced, worldly, and highly respected. Shippen returned to Philadelphia in 1762, probably in the spring. Morgan arrived three years later, deciding to travel throughout Europe after graduation rather than sailing back to America. Morgan traveled to Rotterdam, Antwerp, Paris, Lyons, Rome, Florence, and, finally, Padua, where he visited famed anatomist Giovanni Battista Morgagni, the father of pathologic anatomy, who even into his old age continued to teach anatomy. Morgan was granted admission to some of the most distinguished medical societies in Europe, including the Royal Academy of Surgery in Paris, the Royal College of Physicians of London, and the Royal College of Physicians of Edinburgh. He became the first American fellow of the Edinburgh college. Shippen would become the second.

When Shippen returned to Philadelphia, he began a series of public lectures on anatomy, the first such series in America. He was the first anatomist of any kind to bring true Hunterian methods to the colonies, giving scores of medical students a valuable way of learning a fundamental subject. An obstetrician, Shippen also started a course on midwifery and opened it to men and women alike, a pioneering approach that didn't align with the view of many

Philadelphians, who found males working intimately with females intolerable. The most critical method Shippen brought to the colonies, though, was hands-on dissection of human bodies, the core of the Hunterian method.

Shippen started advertising anatomy courses in the fall of 1762 in Ben Franklin's *Pennsylvania Gazette* and continued advertising each year. He charged as much as six pistoles to attend roughly sixty lectures, depending, most likely, on how many cadavers he could secure during the winter, when classes were held. Students would learn human anatomy, his ads said, as "demonstrated on the fresh subject." Shippen's transparency in the ads about what he was doing—anatomizing a "fresh subject"—was a startling departure from his mentor's approach. Where Hunter went out of his way to ensure that the public never learned how he obtained his subjects, and then ignoring protests when they did, here was Shippen almost bragging about it.

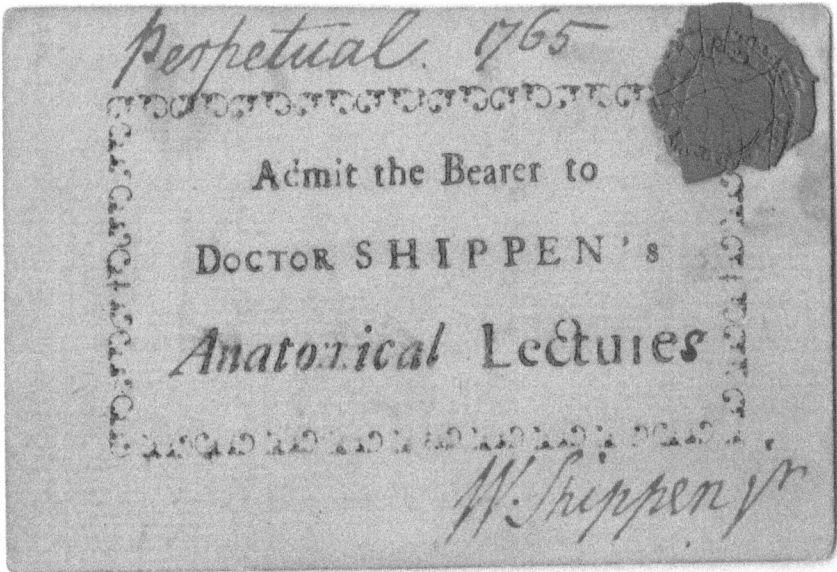

Ticket to attend an anatomy course taught by William Shippen. COURTESY OF UNIVERSITY ARCHIVES AND RECORDS CENTER, UNIVERSITY OF PENNSYLVANIA.

During the fall of 1765, a group of sailors stormed Shippen's house as a protest against his teaching. "For the city being small, almost everyone knew what was going on in it," said William E. Horner, former dean of the medical department at the University of Pennsylvania and author of the first pathology textbook published in America. "The house was frequently stoned, and the windows broken."

That fall, sailors pelted a carriage in front of the house with rocks and fired a musket ball through it. The coachman drove off as fast as he could, barely escaping with his life. The sailors then attacked the home itself, but Shippen had already escaped. Sailors had been hired previously in Philadelphia to disrupt the workings of government, including for what came to be known as the "Bloody Election of 1742." A member of one of the political parties of the time hired dozens of itinerant sailors to riot at the Philadelphia courthouse to disrupt a hotly contested election of assembly members. Whether sailors who attacked Shippen's house had been hired for that purpose has been suspected but not confirmed.

In any event, the rioters left Shippen's anatomy lab alone. "In Philadelphia at that time," explained Mark Frazier Lloyd, archivist emeritus at the University of Pennsylvania, "many of the first-generation residences had long back yards that stretched all the way to the midpoint in the block. Shippen's was one of those. The lab at which he conducted his anatomy classes was back about two hundred feet behind the house." Rioters apparently didn't know where the lab was and so dispersed after searching the main house. The riot became known as the "Sailor's Mob" and was the first recorded instance on this side of the Atlantic of people rioting against a physician for stealing bodies.

Shippen continued using cadavers for hands-on study and defended himself to critics publicly. Aware of the rumors swirling about his practice, he published, in the September 26 issue of the *Pennsylvania Gazette*, just days before the Sailor's Riot, a notice informing the public that rumors about him stealing bodies were mistaken. "It has given Dr. Shippen much pain to hear, that not-withstanding all the caution and care he has taken to preserve the utmost decency in opening and dissecting dead bodies, some evil-minded persons . . . have reported . . . that he has taken up some persons who were buried in the church burying ground." Shippen went on to say that such reports were "absolutely false" and that his dissection subjects had either "wilfully murdered themselves or were publickly executed, except now and then one from the potter's field." Shippen was neither the first nor the last anatomist to declare his innocence in the tawdry trade of body snatching.

William Shippen had no idea what John Morgan would do when he returned to Philadelphia. If he thought Morgan would contact him so they could continue their collaboration in setting up the first med-ical school in the colonies, he was heartbreakingly mistaken. When Morgan traveled home to the city in 1765, he almost immediately met with trustees of the College of Philadelphia and presented, on his own, a proposal to open a new medical school with himself as its head. The trustees approved the plan at the same meeting, on May 3, and, without discussion, created the first medical school in America, now part of the University of Pennsylvania. They also, at that very same meeting, named Morgan as professor of medicine, head of the new department, critical decisions made at dizzying speed.

"It has always seemed to me unlikely," said Lloyd, "that Morgan could come back from England and Europe, and within a few weeks walk into a trustees' meeting, submit an idea for a whole new school, and have it approved on the spot unless it was orchestrated by some-body." That somebody, Lloyd believes, was William Smith, the bril-liant professor of ethics at the college and its first and longest-serving provost. Born in Aberdeen, Scotland, Smith was elected provost in 1755 and served continuously until 1779, save for the war years, and then served as a trustee until 1791, a remarkably long period in posi-tions of great power.

Morgan had taken classes with Smith and had written his professor a note of gratitude while he was away at war. Smith replied warmly that Morgan's expressions of gratitude were but a confirmation of the "goodness of your heart." He then quickly added, "My services to you have not been so disinterested as you imagine. I have a higher demand on you than any pecuniary return which your present or future fortunes will ever enable you to give."

So when Morgan came knocking at Smith's door, it only too readily opened. "Smith was a man of self-aggrandizement his entire life," said Lloyd, "and he would do anything that he thought would advance his own standing, his own career. The moment Morgan told Smith—or told one of his benefactors, who in turn told Smith—the moment that happened, Smith realized that Morgan's proposal was a way to enlarge his own sphere."

With that, Shippen was out. That Morgan had failed to consult Shippen was an appalling affront to Shippen's pride. Shippen might have made discrete inquiries about creating a medical school himself, before Morgan returned, but he would have done so at Pennsylvania Hospital, not the College of Philadelphia. "But the hospital apparently turned him down," said Shippen biographer Stephen Fried. "The managers had no interest in, or money for, broadening their mission beyond the care of indigent patients; and Shippen, a fine young surgeon but not a big medical thinker, did not inspire them to change their minds."

Morgan's slight affected Shippen deeply. Shippen was, after all, the son of William Shippen Sr., a religious and political leader in the city and a well-respected physician in his own right. He had founded the First Presbyterian Church in Philadelphia and was a trustee of the College of Philadelphia during the time of Morgan's proposal. Morgan was a year younger than Shippen and was a member of the first class of the brand-new College of Philadelphia, graduating in 1757 at age fifteen. Shippen, on the other hand, had graduated from the College of New Jersey, now Princeton. Prior to his return to America, Shippen had been recommended to medical leaders in the city by the eminent John Fothergill, who then added in the recommendation that Shippen "will soon be followed by an able assistant, Dr. Morgan." Fothergill assumed, as did others later, that Shippen would lead the team.

Both men were ambitious and confident in their abilities. Their temperaments, though, differed considerably. Historians have variously described Shippen as "calm, cautious, far-seeing, self-possessed, and at times subservient to a crafty nature," as well as "bright, overbearing, and far too sure of himself." Morgan, on the other hand, was considered "impulsive, fervent, positive, and a statesmanlike organizer." Shippen was an eloquent speaker and highly popular; Morgan was neither. Both Shippen and his father had attended the College of New Jersey. Morgan had attended the College of Philadelphia. Morgan worked diligently, if furtively, to ensure that his proposal went to his alma mater, not to Shippen's. Perhaps they were destined to oppose one another.

When the new medical school opened, Morgan was named professor of medicine, the first such position in the nation. Shippen was named professor of anatomy, also the first in the nation, though a solid step down from Morgan. Two weeks later, Shippen wrote to the trustees acknowledging his appointment: "The institution of medical schools in this country has been a favorite object of my attention for seven years past," his letter began, "and it is three years since I proposed the expediency and practicability of teaching medicine in all its branches in this city."

Shippen, still smarting that he had been left out of Morgan's plan, added a biting reminder that he had developed his own plan for a medical school long before Morgan arrived. "I should long since have sought the patronage of the trustees of the college," he wrote, "but waited to be joined by Dr. Morgan, to whom I first communicated my plan in England and who promised to unite with me in every scheme we might think necessary for the execution of so important a point." Ever tactful, Shippen ended with, "I am pleased, however, to hear that you gentlemen, on being applied to by Dr. Morgan, have appointed that gentleman professor of medicine."

Into that clash of powerful personalities strode Benjamin Rush, a young Philadelphia physician who would one day sign the Declaration of Independence and become the most acclaimed physician in the nation. Rush knew both men and respected each of them. Younger than either, Rush learned medicine as an apprentice, as had Shippen. Rush worked under a Philadelphia physician named John Redman, as had Morgan earlier. Redman taught Rush techniques in bloodletting, cupping, and delivering an infant, just as he had taught Morgan before. When Shippen offered his first anatomy course, Rush signed up immediately. Rush soon considered Shippen his "friend and master."

Rush knew about Morgan as well, having lived in the same neighborhood as children, and might even have gone to Morgan's father's store at the corner of Second and Market Streets. Rush also signed up for Morgan's *materia medica* class at the newly opened medical school and almost certainly had a sense of the Shippen versus Morgan ruckus. "As a keen reader of newspapers and people," said Fried, "who had developed a fascination for institutional gossip—he surely knew of the friction between the two professors and perhaps wondered how it might impact his career. In fact, when he elected to take Shippen's anatomy class for a second time that year, he may have done it just to appear he wasn't choosing sides." Rush had also studied medicine at the University of Edinburgh, where he made sure to study chemistry. Morgan had told Rush before Rush left for Scotland that he would hold open a professorship of chemistry for him.

Rush dedicated his graduate thesis to both men but made the blasphemous mistake of placing Shippen's name first and Morgan's second, a slight that incensed Morgan. Rush wrote several letters of apology. "Dr. Shippen was my oldest friend and master, and this was the only reason why I put his name before yours," Rush wrote in a letter to Morgan. "I hope therefore, my good friend, you will overlook my omission in placing your name after Dr. Shippen's. My thesis has been read, and is now I dare say forgot forever. No one will remember six weeks hence whose names were mentioned in my dedication."

Morgan's ego somewhat assuaged, he persuaded the trustees to name Rush, in 1769, as the first professor of chemistry in the school and the nation. As Rush was working to build his practice in Philadelphia, he realized that another of his mentors had turned against him. Shippen had begun telling his students not to take Rush's chemistry course. Said Fried, "Since Shippen left almost no writing, it's hard to know today why he did this. He could have been unhappy that Rush sided with Morgan on the college faculty, especially since Morgan regarded Shippen's specialty, surgery, as inferior to what he practiced and taught. Or, perhaps they just didn't get along. Either way, Rush felt Shippen was becoming his 'enemy.'"

In October 1775, in preparation for war, Morgan was named the second director-general of the military hospitals for the Continental Army, following the ouster of the first director-general, Benjamin Church, who had been dismissed for "treasonous correspondence" and later court-martialed for being a British spy. Morgan took over under impossible conditions. Medicines and medical supplies were scarce, medical personnel even scarcer. Within the military medical department there existed no clear chain of command. Logistical problems seemed insurmountable. Personnel from simple soldiers to experienced surgeons were unhappy with their pay and the frequent delays in receiving it.

The Continental Congress took note of the disorganization and the number of soldiers dying from disease and sought to place blame. Congress had never defined the roles and responsibilities of the medical department or its head. It had never funded the department as well as it should have, nor done anything needed for the director-general to centralize care. The scapegoat became John Morgan, who was fired on January 9, 1777, and replaced with Shippen.

Morgan believed to his dying day that Shippen had orchestrated his removal.

Shippen faced the same obstacles as Morgan, however. "Like Morgan before him," explained historian Lassek, "Shippen was haunted by shortages of essential supplies, by faulty organization and invincible rivalries and jealousies, and by the indifference and incomprehension of Congress. He had all the handicaps Morgan had to labor under—and one more: he did not like to work. The result was that he became the object of a new flood of criticisms of the medical department and, just as surely as Morgan had been, he was destroyed by them."

Morgan might have been gratified that his rival faced the same outcome as he had, but it mattered not. Morgan decided to enlist Rush in a quest for revenge. Rush, his insecurity still smarting from Shippen's turning against him, raised concerns about Shippen's handling of the military hospitals in a letter to John Adams in 1777, saying that Shippen "is both ignorant and negligent of his duty" and that his hospital system is "a mass of corruption and tyranny." Morgan soon urged the Continental Congress to court-martial Shippen. Congress balked, not exactly eager to replace yet another director-general. Over and over, Morgan pushed for court-martial, Rush assisted as best he could, and Shippen defended himself against all charges. Finally, in 1780, Shippen was acquitted of all but one charge and was dismissed from his position but was not court-martialed.

The long, vitriolic conflict between Shippen and Morgan had finally, if only superficially, ended. Their feud had damaged both of their reputations and that of their families as well. It had also fractured the medical school in Philadelphia into Morgan supporters and Shippen supporters, a rift that didn't truly heal until after both men had died.

CHAPTER 6

BULLETS, BLOODLETTING, AND BAYONETS

THE TURMOIL AMONG THREE OF THE COLONIES' MOST IMPORTANT medical figures is emblematic of the turbulent state of American medicine in the eighteenth century. Medical advances came slowly from Europe and sometimes proved difficult to implement. Case in point, the treatment of smallpox outbreaks in the colonies during the first half of the century.

Smallpox was a deadly infection caused by the variola virus. Victims suffered high fever, nausea, vomiting, and a distinctive rash with pustules. Severe cases involved kidney failure, lung damage, severe blood loss, and death. Inoculations against the disease had been used throughout Europe, China, and parts of Africa since at least the 1400s. Inoculating someone consisted of transferring pus from someone's smallpox sore to an open wound on someone else, often a small incision in the arm of the person being inoculated. Virus in the pus would infect the person and typically cause mild disease, prompting the body to develop antibodies. The antibodies didn't prevent infection, but they did ensure that the person didn't become as ill as they might have.

Smallpox inoculations, or variolations, weren't often used in America at that time. Clergy believed that "smallpox was God's way to punish sinful people," explained Per-Olof Hasselgren, professor of surgery at Harvard University, "and trying to prevent the malady was to interfere in God's plans." Boston hadn't seen a case of smallpox for nineteen years when, in 1721, a ship from the Dry Tortugas sailed into its port with a crew member infected with the virus.

Puritan minister Cotton Mather had heard about smallpox inoculation through his enslaved African, Onesimus, who had been variolated when he lived in Africa. Mather spoke with Boston's fourteen doctors about the process, trying to persuade them to start inoculating the citizenry. Medical historian Francis Packard said that doctors "ridiculed the idea and treated the proposition with scorn." Although religious objections formed part of the resistance to inoculations, it was by no means the only one. Most physicians in the colonies had been trained through an apprenticeship model and lacked the kind of formal education Europe's medical universities could supply. One physician who had been trained in Europe, William Douglass, balked at inoculation because he considered the procedure "quackery that promised to do more harm than good to public health." Regardless of their reasoning, all of the doctors Mather spoke with turned him down, with one exception: his friend, Zabdiel Boylston, who, said Packard, "at once saw the value of the remedy and entered eagerly into the scheme to stamp out the plague."

Boylston began inoculating residents but ran headlong into public condemnation. "Indeed, so strong was the opposition to inoculation," said Hasselgren, "that Boylston had to go into hiding. On one occasion, his wife and children were threatened by a hand-grenade thrown into their home." Not only did doctors in the city oppose inoculations, but so too did newspapers and the courts. "Many pious, respectable personages," said Packard, "were of the opinion that should any one of [Boylston's] patients die the doctor should be hung for murder."

Mather, too, felt the public's wrath. In November 1721, someone tossed a small bomb through a window in his home, landing in a room where Mather's nephew was recovering from his own inoculation. The fuse burned out before the device could explode, and a note attached to it remained intact. The note read, "Cotton Mather, You Dog, Dam you, I'l [sic] inoculate you with this, with a Pox to you."

In spite of the threats, Boylston succeeded in inoculating 276 people, including his own son, a Black servant named Jack, and Jack's two-and-a-half-year-old son. Of the 276 inoculated, only 6 died of the disease, a remarkable success. Mather and Boylston's work paved the way for widespread inoculations in Boston when another major outbreak struck in 1764.

Boylston was not formally trained as a physician; he learned medicine from his father. His success with inoculations would provide him a trip to England and a fellowship in the Royal Society of London. Like Boylston most doctors practicing in the colonies started their careers as apprentices, learning at the hands of a senior physician, who probably also learned through an apprenticeship. That model of training led to a wide range of experience and expertise in its practitioners. A family of physicians in Marlboro, Massachusetts, provides an illuminating example of medical apprenticeships at the time. John Gott, a tanner, had three sons, two of whom followed in his footsteps in the tanning business and one, Benjamin, who became indentured to Samuel Wallis, a physician in Ipswich. John sent Benjamin, probably around age twelve or thirteen, to Wallis to learn the "art and mysteries" of becoming a doctor. Wallis agreed to pay Benjamin "£200 in silver money or in good bills of credit when he arrives at the age of twenty-one years."

Benjamin finished his indenture, received his payment, and moved to Marlboro in 1727 to practice his trade. In early 1733, Benjamin accepted one Hollister Baker, about sixteen years old, into an apprenticeship of five years and four months. Baker's parents promised Benjamin that "his master's goods he shall not waste, embezel, purloine or lend unto others, nor suffer the same to be wasted or purloined. . . . Taverns nor alehouses he shall not frequent; at cards, dice, or any other unlawful game he shall not play; fornication he shall not commit, nor matrimony contract with any person, during said term: from his Master's service he shall not at any time unlawfully absent himself but in all things as a good, honest and faithful servant and apprentice, shall bear and behave himself."

Gott, in turn, promised to "teach the said apprentice, or cause him to be taught by the best ways and means that he may or can, the trade, art or mystery of a physician according to his own best skil and judgm't (if said apprentice be capable to learn) and will find and provide for and unto said apprentice, good and sufficient meat, drink, washing, and lodging during said term both in sickness and in health."

Baker performed a variety of functions for his preceptor. He "ran his master's errands, washed bottles, mixed drugs, spread plasters, and finally, as the stipulated term drew towards its close, actually took part in the daily practice of his preceptor—bleeding his patients, pulling their teeth, and obeying a hurried summons in the night." Apprentices learned how to set broken bones, treat sores and wounds, excise harelips, lance abscesses, and prescribe salves, creams, poultices, and other topical therapies that they also sold in their office. They learned those skills not in a classroom but in the field and not from a professor accustomed to the ways of teaching but from someone who had also learned his skills through an apprenticeship. Only about 10 percent of an estimated 3,500 physicians in America at the start of the Revolutionary War possessed legitimate medical degrees, and nearly all of those were trained in Europe. Medical journals didn't exist, and textbooks were scarce.

Apprenticeships without oversight and quality control led to practitioners with inconsistent abilities, possibly poor decision-making, and sometimes outright quackery. Politician and historian William Smith wrote about quackery in the colonies, saying in 1755 that "quacks abound like locusts in Egypt" and that "any man at his pleasure sets up for physician, apothecary, and chirurgeon." William Livingston, a lawyer and the first governor of New Jersey, railed about the wide disparity in doctors' skills, writing in his weekly newsletter, "Independent Reflector," in 1753, "No man is of greater service or detriment to society than a physician. If he is skillful, industrious, and honest, he is of unspeakable benefit to mankind; but if incapacity, idleness, and roguery are his characteristics, he is a curse to the community. There is no city in the world, not larger than ours [New York], that abounds with so many doctors . . . the greatest part of them are mere pretenders to a profession of which they are entirely ignorant."

Livingston proposed a nine-point plan to the New York legislature that would require all physicians, surgeons, and apothecaries to become licensed by passing an examination, punish those who practice without a license, and conduct yearly inspections of any shop that sells drugs. The proposal formed the basis of a law passed in 1760 that would regulate the practice of medicine. That law, however, was limited to New York City, not the entire state.

Virginia had taken a swipe in 1639 of regulating medicine with a law controlling the cost of medical care. The act allowed courts to adjudicate whether a doctor's or surgeon's rates were "immoderate and excessive." The act did nothing, though, to regulate actual medical practices, nor did it limit prices. Doctors and quacks alike continued to charge whatever they wanted, and the legislature seemed powerless to stop it.

In Massachusetts, along with most other colonies, medical care was essentially a free-for-all throughout the 1700s. Reginald H. Fitz, a prominent Harvard pathologist, said in 1894:

> In this state anyone who chooses may practice medicine. He has but to announce himself a physician and he becomes one. He may assume a title to which he has no claim, and may place a forged certificate upon his walls. He may advertise himself a graduate of any institution he prefers; may claim to have accomplished any number of cures of what have been pronounced incurable disease, he may promise preventives and specifics against any and all maladies; he may publicly announce the most glaring untruths—all for the sake of deceiving and fleecing a credulous public—and the law cannot interfere with his actions.

The word "doctor," then, covered a variety of practitioners working in all sorts of environments. One Philadelphia wigmaker advertised that his shop had an "experienced doctor" who could "bleed, draw teeth, and cure all manner of wounds incomparably well." Even shops without doctors sold medical goods, giving the general public a diverse and highly competitive marketplace for their medicines, all of which left the consumer as the first and often only arbiter of their own care. Nissa Strottman, an attorney who studied regulations on medical practice in colonial America, particularly those in Philadelphia, described how often citizens sought medical care from a doctor:

> *Everyone except the extremely poor, who relied on poor relief for medical care, and servants and the enslaved, could directly consult a doctor or another practitioner. However, a patient's wealth determined how practitioners were used. For instance, those who had little money called on doctors only for serious cases, and generally did not see doctors often unless their illness required repeated visits. Those who could afford to see a doctor frequently did so, often running up rather large medical bills in the process. People who either could not afford to see a doctor, or did not wish to, had other medical alternatives.*

Those medical alternatives included patent medicines advertised as "cures" for whatever the ailment might be. Ben Franklin's mother-in-law, Sarah White Read, got into the action with her own salves and ointments, which Ben advertised in his *Pennsylvania Gazette*. "The Widow Read," said one advertisement, "continues to make and sell her well-known ointment for the itch, with which she has cured [an] abundance of people in and about this city for many years past. It also kills or drives away all sorts of lice in once or twice using." Numerous patent medicines from England sold widely in the colonies, including Hooper's Female Pills, Bateman's Pectoral Drops, Godfrey's Cordial, Turlington's Balsam of Life, and Steer's Opodeldoc, a concoction made of camphor, rosemary oil, and soap. Many other such medicines were created by colonists, each with their own mixtures of herbs, oils, barks, and other plant or animal products.

Some homemade mixtures actually worked for certain conditions, such as Williams' Pink Pills for Pale People, which contained iron and successfully treated anemia. Other mixtures proved completely ineffective, and some of them, such as those containing opium

Bottle of Turlington's Balsam of Life with the 1754 design, embossed with "By the King's Royal Patent Granted to Robt Turlington for His Invented Balsam of Life. London." COURTESY OF JEREMY KEMP, THE ORIGINAL AND GENUINE CURES ALL DISEASES.COM.

or mercury, proved addictive or deadly. One plant, *Atropa belladonna*, or deadly nightshade, had been used in medicines since ancient times. Nightshade contains atropine, which increases the heart rate but can also cause tremors, delirium, coma, and death. In small doses, it was said to alleviate whooping cough, scarlet fever, spasmodic asthma, and intestinal cramps, but higher doses proved fatal.

With so many medicines being sold by so many people with so little medical background, competitive tensions and professional jealousies were bound to erupt. One contentious debate erupted in Virginia in the 1730s between John Tennant and most of his colleagues about the value of a particular herb. Tennant called himself a "Practitioner in Medicine," though whether he obtained actual medical training is doubtful. While in London in 1737, he persuaded three physicians to attest to his qualifications for a degree in physic from the University of Edinburgh and eventually failed to obtain the degree. On his return to Virginia, he proudly flaunted letters from the physicians, attesting to his abilities, and continued to call himself a medical practitioner.

Historian Richard M. Jellison wrote about Tennant: "A prolific writer, he was well known among Virginians, many of whom considered him a benefactor to the colony. Yet, most of his professional colleagues resented his contumacy and believed that he was a charlatan." Tennant believed completely in the power of an herb called Seneca rattlesnake root to cure pleurisy, a catch-all term at the time that generally referred to pneumonia, flu, or certain other respiratory conditions that, together, caused a staggering number of deaths in colonial Virginia. Tennant argued his view on rattlesnake root so often and so loudly that Virginia physicians, who didn't believe in the herb, fought back. One calling himself "I. C." tried to correct Tennant's "false suggestions," writing that "to make a seemingly dying woman, in four hours after first taking of it, rise from her bed, and be able to whip her children [was about as probable] as if it was said the root will raise the dead to life."

Although the herb initially failed to catch on, it would become, over the years, a staple of nineteenth-century medicine and was used for cough, asthma, pneumonia, whooping cough, and gout. Tennant's belief in Seneca snakeroot, said Jellison, "cannot be questioned. His decision not to keep his discovery secret for personal aggrandizement is admirable. The later acceptance and widespread use of the Seneca root in American medicine seemed to justify his claims."

Colonial medical care sometimes bordered on the bizarre. Cotton Mather once wrote about a woman who had been treated for a painful abdominal condition. "The wife of Joseph Meader," wrote Mather, "had long been afflicted with that miserable distemper known as the twisting of the guts. Her physician advised her to swallow a couple of leaden bullets; upon which after some time, her pain was abated and the use of her limbs returned to her." Mather added a perhaps unnecessary codicil: "Attempts to swallow bullets have not always terminated so well."

A substance in widespread favor from the Middle Ages into the 1900s was called by one seventeenth-century physician "one of the greatest improvements ever made in practical medicine." It was said to treat pneumonia and rheumatism, as well as dysentery, diseases of the liver, hydrocephalus, smallpox, tuberculosis, typhus, and yellow fever. Exactly how it helped all of those conditions wasn't clear. One physician claimed that the substance transferred "diseases of the head, of the eyes, and of the bowels to the mouth, where they are less dangerous and more manageable." Another explained that the heavy weight of the element cleared "morbid matter" from the digestive tract, thereby "promoting better circulation and eliminating disease." That magical element was mercury, known now as extremely toxic. Mercury can cause nosebleeds, diarrhea, loose teeth, kidney disorders, headaches, tremors, and blindness. Its use in medicines has been phased out almost completely worldwide.

The most common medical treatments in colonial America, though, were purging, vomiting, blistering, and bloodletting, a particularly dangerous practice in almost universal use since the Egyptians tried it 3,000 years ago. The earliest medical men believed that the healthy body consisted of a balance of four "humors": blood, phlegm, black bile, and yellow bile. An excess or deficit in a humor was associated with particular diseases or personality characteristics. Someone who was considered as sluggish or dull, for instance, was thought to have an unemotional, or "phlegmatic," personality. Someone with a temper possessed an excess of yellow bile and was called "choleric." To treat imbalanced humors, the patient was bled, with the practitioner draining significant amounts of blood from a vein. Bloodletting, in the words of an anonymous twelfth-century writer, "makes the mind sincere, it aids the memory, it purges the brain, it reforms the bladder, it warms the marrow, it opens the hearing, it checks tears, it removes nausea, it benefits the stomach, it invites digestion, it evokes the voice, it builds up the sense, it moves the bowels, it enriches sleep, it removes anxiety."

The procedure consisted of using a sharp instrument to pierce a vein below the skin and allow blood to flow freely out, typically into a bleeding bowl, a wide, shallow bowl with a handle. Initially practitioners used thumb lancets to cut into the patient's flesh. Thumb lancets were small, double-edged knives with a horn or shell handle, often encased in an ivory or tortoiseshell case, convenient for carrying in a pocket. Later devices called fleams came into widespread use. A fleam looked not unlike a Swiss Army knife, with at least one and often three or more pointed, sharp-edged blades folded together into a case, each blade having a different-sized tip. Spring-loaded fleams also came into use. They operated by pushing a lever or button on the handle, which would spring open a blade and slice the skin. Spring-loaded fleams made more consistent incisions but were more difficult to clean, thus causing more infections.

Leeches were also used in bloodletting, though mostly for children, the elderly, or terribly weak patients. Leeches would be applied to the skin and allowed to bite into the flesh. Because leeches leave behind a mild anticoagulant when removed, as much as two ounces of blood could be removed during one application.

The amount of blood removed during each bloodletting varied widely. An initial bloodletting could pull twenty ounces of blood from the patient, about as much as a venti coffee from Starbucks. Sometimes the practitioner would continue to drain blood until the patient fainted. Repeated bleedings were often performed, in some cases removing half of the person's entire blood volume of about 200 ounces. In contrast, today's blood donations consist of removing

A bleeding bowl gifted to Queen Mary by Middlesex Hospital in 1929. Doctors would examine blood caught in the bowl for clues about the state of the body.
COURTESY OF ROYAL COLLECTION TRUST/© HIS MAJESTY KING CHARLES III, 2024.

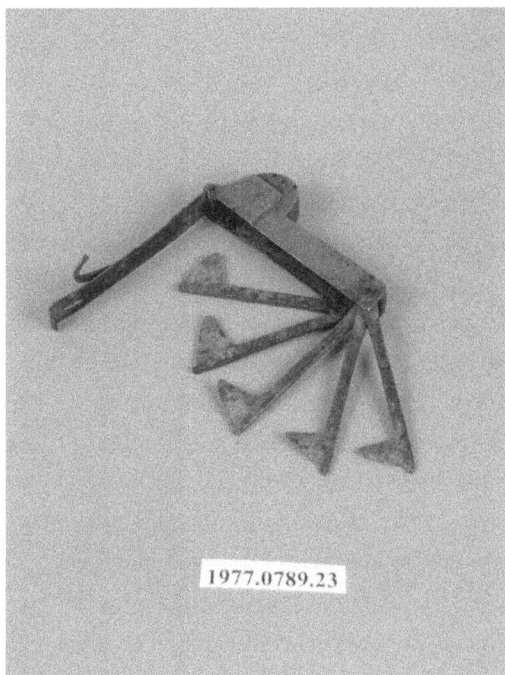

Fleam with five blades.
COURTESY OF THE DIVISION OF MEDICINE AND SCIENCE, NATIONAL MUSEUM OF AMERICAN HISTORY, SMITHSONIAN INSTITUTION.

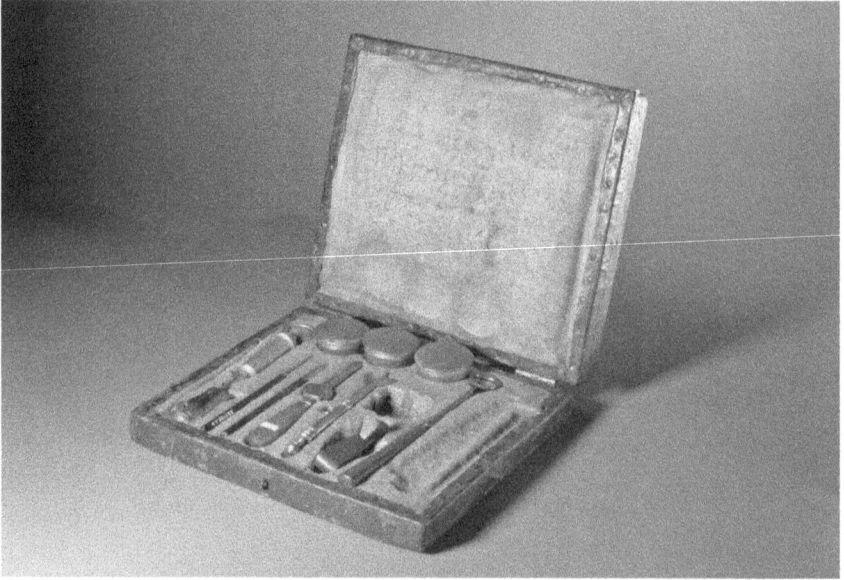

Set of bloodletting instruments. COURTESY OF THE DIVISION OF MEDICINE AND SCIENCE, NATIONAL MUSEUM OF AMERICAN HISTORY, SMITHSONIAN INSTITUTION.

less than sixteen ounces of blood once every eight weeks. Although today's phlebotomy has proven extremely safe, bloodletting procedures in earlier centuries proved consistently perilous. Among the more famous patients who underwent bloodletting and subsequently died were King Charles II, Wolfgang Mozart, and George Washington.

Following a seizure and "the dreadfulest shriek," King Charles II's physicians "bled him sixteen ounces. They cupped him, scarified him, and blistered him with cantharides; they gave him emetics, cathartics, and enemas. They bled him again, eight ounces. All before noon." Physicians bled the king again the next day and also the next. They treated him with a wide array of plant and animal substances, including cream of tartar, mallow root, white wine, chamomile, nutmeg, Peruvian bark, and something called "spirit of human skull." It was all for naught, and the king died on February 6, 1685.

Mozart's doctors couldn't save him with bloodletting either. The composer had been ill for about two weeks from a stubborn infection and had been suffering from an extremely high fever, vomiting, and rash. His physician performed at least one and possibly three bloodlettings in the last days of his life. "When the doctor was called in," recalled Mozart's sister-in-law, Sophie Heibel, "he bled Mozart and put cold compresses on his burning head, whereupon his strength declined rapidly and he fell unconscious and never came around again." Mozart died at age thirty-five on December 5, 1791. The cause of the great composer's death has never been fully determined, but most medical historians believe that heart failure, acute kidney disease, and severe dehydration were the dominant factors and that, in the face of severe dehydration, bloodletting unquestionably hastened his death.

President Washington didn't fare any better. Washington had been overseeing his farm on horseback on December 12, 1799, when it began to snow. The snow changed to hail and then to rain, leaving the former president's riding clothes soaked through by the time he reached home. He awoke the next morning with a sore throat and again went out riding, this time in a new snow. He felt sick on his return, and over the next several hours, he continued to worsen. Doctors were called. Washington received an enema and underwent four bloodlettings, which removed about five pints of blood, half of his entire blood volume. At dinner time on December 14, he told his dear friend and physician, James Craik, "Doctor, I die hard, but I am not afraid to go." Washington died a few hours later.

Bloodletting was used during the Revolutionary War to treat several types of traumatic injuries, even those that had already caused significant blood loss. John Jones, one of the founders of the medical school at King's College and designer of the wards at New York Hospital, published, just in time for the war, the New World's first medical and surgical manual. In it, he recommended bleeding for most gunshot wounds, despite also recognizing its dangers. Famed Philadelphia physician Benjamin Rush was a staunch supporter of bloodletting and came to believe that there existed just a single disease in the world. He further believed that this lone disease should be treated solely, said historian Mary C. Gillett, by "a low diet, vigorous purges with calomel and jalap, and bleeding until the patient fainted."

Gillett went on to explain that "Rush apparently did not hesitate to remove a quart of blood at a time, or, should unfavorable symptoms continue, to repeat such a bleeding two or three times within a two- to three-day period, it being permissible in his opinion to drain as much as four-fifths of the body's total blood supply. In time, Rush's system and treatment became, in the words of a noted medical historian and physician, 'the most popular and also the most dangerous system in America.'"

Rush favored bloodletting so much that he would, in 1797, sue a journalist named William Cobbett for libel. Cobbett had reviewed mortality records for Philadelphia during the yellow fever epidemic of 1793 and determined that mortality rates "increased significantly" after bloodlettings and "mercury purges" from Rush. "Dr. Rush, in that emphatical style which is peculiar to himself," wrote Cobbett, "calls mercury the Samson of medicine. In his hands and those of his partisans it may indeed be justly compared to Samson: for I verily believe they have slain more Americans with it than ever Samson slew of the Philistines. The Israelite slew his thousands, but the Rushites have slain their tens of thousands." Rush sued and, after long delays, finally triumphed. The court fined Cobbett $5,000 for publishing his accusations, accurate or otherwise.

Bloodletting, at long last, began to lose favor during the 1800s. A Paris physician named Pierre-Charles-Alexandre Louis carried out one of the first clinical studies in history about the effectiveness of bloodletting. Louis, considered one of the first medical epidemiologists, studied bloodletting in seventy-seven patients with pneumonia. He concluded that his results "establish narrow limits to the utility of this mode of treatment." He stopped well short of decrying the practice, however, writing that bloodletting "should not be neglected in inflammations which are severe and are seated in an important organ." Gradually, as more clinical studies examined bloodletting, the medical community distanced itself even more from the practice until the mid-1900s, when it all but disappeared.

Like bloodletting, surgery during the eighteenth century was fraught with danger. Operations tended to be hectic affairs, with surgeons moving quickly to reduce the amount of pain experienced by the patient. Surgeons lacked any kind of reliable anesthetic. Alcohol was commonly used and sufficed to dull the person's pain but not prevent it. Renowned Harvard surgeon Henry J. Bigelow wrote in a treatise on the history of anesthesia, "Patients, while dead drunk, have been operated upon painlessly, and a dislocated hip was thus reduced after a bottle of port wine." Ancient Egyptians often mixed wine with mandrake, and belladonna, cannabis, and opium were also used as anesthetics. A thirteenth-century monk named Theodoric used a sponge boiled with a mixture of hemlock, henbane, lettuce, mandrake, mulberry, and opium as an anesthetic and then had the patient breathe the vapors. It wasn't until ether caught on in the mid-1800s that Americans had any hope of undergoing surgery without intense pain.

Ether, invented in 1540, was made at the time by "distilling sulfuric acid (oil of vitriol) with fortified wine to make an *oleum vitrioli dulce* (sweet oil of vitriol)." The mixture is extremely flammable, however, and rather prone to explosion. Ether did not find widespread use in medicine until the nineteenth century. Before that, the main use was as a recreational drug among poor Britons when they didn't have access to alcohol. "American students adopted a variation of this practice," said Connie Y. Chang, an anesthesiologist at Rutgers New Jersey Medical School, "in the 'ether frolics' of the early 1800s to achieve a feeling of euphoria. Participants would hold ether-soaked towels to their faces until losing consciousness."

Surgeons in early America contended with crude, dirty instruments as well. Joseph Lister wouldn't employ antiseptics to sanitize instruments until the 1860s, and until then, wound infections were common and often fatal. Physicist Daniel Fahrenheit had only recently, in 1724, invented a scale for mercury thermometers that could help physicians gauge temperature accurately, though few used it, and French physician René Laennec wouldn't invent his stethoscope until 1816. Until the arrival of Louis Pasteur's germ theory in 1861, stating that viruses, bacteria, and other microscopic organisms caused disease rather than imbalanced humors, surgery remained a dangerous option. "The conditions under which the surgeons toiled," said historian Richard Blanco, "their primitive instruments, the lack of adequate lighting, their ignorance of the cavities of the body, the filthy circumstances of their operating wards, made surgery a very grim business."

American surgeons also lacked textual resources to aid them in surgical techniques. At the start of the Revolutionary War, almost all medical books in the colonies had been brought from Europe, and there were scarce numbers of them. Just one book had been published by an American author, the surgical manual by John Jones, and American physicians had authored only about twenty pamphlets. Army surgeon John Billings, writing in *A Century of American Medicine*, explained that "the only public medical library was that of the Pennsylvania Hospital, which contained, perhaps, two hundred and fifty volumes. There were probably not two hundred graduates of medicine in the country, and not over three hundred and fifty practitioners of medicine who had received a liberal education."

Books for the military were in short supply as well. *Observations on the Diseases of the Army*, published in 1752 by Edinburgh physician Sir John Pringle, proved helpful in combatting illness in encampments, particularly typhus, a deadly bacterial infection. John Ranby, a British surgeon, published a manual for treating gunshot victims in 1744 that made its way to the colonies. A quick-reference manual by Baron Gerhard Von Swieten, called *Swieten's Diseases Incident to Armies*, was available as well and was apparently aimed at untrained individuals caring for the sick and injured when no trained individuals were around. Whether the typical surgeon possessed enough background knowledge and clinical wherewithal to make effective use of the books is another issue entirely.

Most surgeons during the Revolutionary War came to their posts having been apothecaries or medical apprentices. They might have been honest and well-intentioned, perhaps even competent, but most never would have seen the kinds of traumatic injuries they would see during the war. Wounds from gunshots, bayonets, and cannonballs were commonplace at the frontlines, all while such diseases as dysentery and smallpox ran rampant through camps. Military surgeons treated sprains and strains, broken bones, dislocated joints, burns, head injuries, bayonet and gunshot wounds, and limbs ripped off from cannonballs. Perhaps the most painful and dangerous operation during wartime was the amputation. New York surgeon Frederick S. Dennis described in 1929 what amputations during the Revolutionary War were like:

It is only necessary to remember that a little over a hundred years ago there were scenes enacted in the name of surgery which eclipsed in horror the frightful cruelty of the Spanish inquisition, the untold miseries of the Bastille, the indescribable sufferings of the Black Hole of Calcutta, and the excruciating pains of the Turkish bastinado and the cruel massacre of the Huguenots. Patients were held down upon the operating table by brute force and were operated upon while in the full possession of their senses; they were heard to cry out in heart-rending screams for a discontinuance of the tortures.

Even given Dennis's melodramatic phrasing, his use of the word *torture* wasn't far off. The process, without anesthesia, was hellacious. To limit pain, surgeons learned to operate as quickly as possible. If the patient was an officer, rum, brandy, or any other available alcohol might be given. Non-officers, though, would typically be given a piece of wood covered with cloth to be gripped between their teeth.

Surgeon's mates would restrain the patient, lying on his back on a table. The surgeon would wrap a tourniquet, usually made of leather, a hands-breadth above the location of the amputation, and

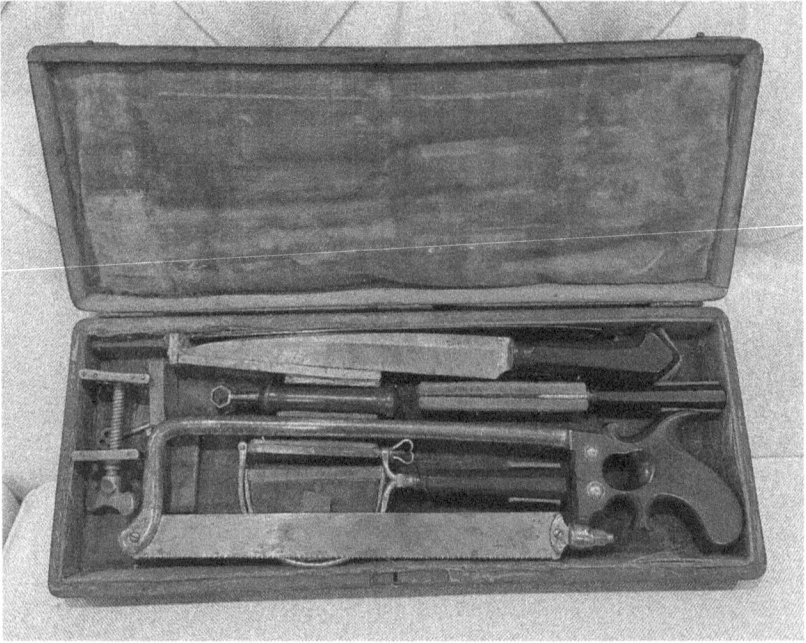

Typical amputation kit with knife, hacksaw, and other implements.
COURTESY OF THE ARBITTIER MUSEUM OF MEDICAL HISTORY.

twist it tight using a wooden dowel. Surgeons could also use a Petit screw tourniquet, a device that uses a screw to gradually tighten or loosen a belt around the extremity. Tourniquets greatly reduce blood flow and help prevent major blood loss during an operation. After the tourniquet is applied, the surgeon would use a long, curved knife to slice through the skin, fat, and muscle in a circle around the limb. He would then push back the tissue and use a saw to cut through the exposed bone. In a leg amputation, he would clamp the femoral artery and other large blood vessels, and then tie the end with a silk or linen suture to reduce bleeding when the tourniquet was removed. The skin and underlying tissues would then be sutured closed and the stump bandaged in white linen.

Little information exists about the number of amputations performed during the Revolutionary War, but during the Civil War, Army surgeons used 13,000 Petit screw tourniquets and 50,000 strap tourniquets to perform more than 29,000 amputations. Each patient undergoing an amputation stood about a one-in-four chance of survival, with most dying from blood loss or postoperative infection.

The Revolutionary War turned simple apothecaries into experienced surgeons and experienced surgeons into medical leaders. It demonstrated the need to create more medical schools, admit more students into medical training programs, and develop better-trained physicians and surgeons for a population that increased from roughly 2.5 million at the end of the war to nearly nine million by 1790. And all of those new students would need access to an anatomy lab and a fresh body for study.

Chapter 7

RESURRECTION, DECAY, AND DISSECTION

The board of governors for New York Hospital met on March 8, 1785, the board's first meeting since the war started. The group convened at the historic Merchants Coffee House at the corner of Wall and Water Streets, a major gathering place and auction house. There the board elected officers for the coming year. They also voted to allow recent immigrants from Scotland to use the hospital for temporary living quarters while they located more permanent housing and approved a request from Richard Bayley and his young colleague, Wright Post, to use one or two rooms in the building for anatomy lectures.

Born in 1766, Wright Post began studying medicine under Bayley's tutelage at age fifteen and, after a four-year apprenticeship, moved to London to study with a well-known anatomist named John Sheldon. Post lived with Sheldon while he attended lectures, a common practice at the time, and worked two years at London Hospital before returning to New York in 1786 at twenty years old. Post was described as being "tall, handsome, and of fashionable exterior, wore long whiskers and his hair powdered and tied back in a queue."

Post's teacher in London had, like Bayley, studied with a famous Hunter brother, John, who taught his students to plunder cemeteries for bodies, just as his brother William did. John, however, proved more ruthless than William in his quests for subjects, famously purloining the body of the "Irish Giant," Charles Byrne. Byrne stood about seven feet, ten inches tall, the result of an undiagnosed tumor in his pituitary gland. He made a living from appearing in freak shows throughout the United Kingdom and couldn't imagine a fate worse than being displayed for public amusement in death, as well as he had in life. His ardent desire was to be buried at sea so that his body would never be stolen, cut open, and displayed. When he died in 1783 at age twenty-two, he would not get his wish.

On his death, wrote one newspaper, a "whole tribe of surgeons put in a claim for the poor departed Irish giant and surrounded his house, just as Greenland harpooners would an enormous whale." Arrangements had been made previously to ship Byrne's coffin to Margate, a town on the southeastern coast of England, and then ferried out to the North Sea, where it would slip silently into the water. John Hunter, however, would have none of it. He is said to have paid £500 for someone to take the man's body from its coffin before it could be shipped to Margate and replace it with weights. With that task completed, Hunter anatomized the body for his students. Four years later, Byrne's skeleton was placed on display in Hunter's own specimen collection, at his home at Leicester Square, before being transferred to the Royal College of Surgeons of England's Museum in 1813, where it remained in a prominent display case until 2017.

The method Hunter used to steal Byrne's body proved atypical; the vast majority of his bodies were stolen by resurrectionists. A student of Hunter's, in a letter to his sister, described an evening in the Hunter household this way: "There is a dead carcass just at this moment rumbling up the stairs, and the Resurrection Men swearing most terribly."

Resurrectionists would monitor cemeteries for burials, which generally occurred during the day. Potter's fields and Black burial grounds proved easiest to plunder; relatives were too poor to pay a watchman to guard the grave, and coffins were usually built of inexpensive wood, making them a cinch to break open. Rural church cemeteries made excellent targets as well, but if the resurrectionists—whether professional, physician, or student—were desperate enough, any cemetery or burial spot would work just fine. Experienced resurrectionists built a network of gravediggers, undertakers, and physicians who would alert them of impending burials.

The ideal time for resurrecting a body was the first night after burial, when the grass around the grave had already been trampled during the funeral and further trampling from gravediggers wouldn't attract undue attention. Rain made removal of the body difficult, and heavy rain would prevent the men from returning the gravesite to its original condition. Snow was likewise problematic unless enough snow would be falling later to cover footprints and carriage tracks.

Anatomy students sent on their own by their teachers might have been less aware of those kinds of tricks, but they stole bodies nonetheless. Sometimes even the professors ventured out at night in the service of their profession. No matter who was doing the pilfering, though, removing a body from a grave could challenge the strongest of stomachs. Author and historian James Breeden explained the basic setup for body snatching:

> This act called for three men and a wagon. One man, after depositing the other two near the cemetery, was charged with returning at a specified time. (What could be more suspicious than a wagon waiting outside a graveyard at night!) The second and third men quickly located the grave. Their first act was to examine it with a carefully shaded lantern for any special arrangements of stones, sticks, or flowers placed there by the family to discourage or detect disinterment. Should anything of the sort be discovered, it was painstakingly mapped in order that it could be duplicated after the grave had been robbed. A large tarpaulin was then spread adjacent to the grave, and on it the excavated soil was deposited, insuring that no telltale dirt was left lying around. Next, a three-foot-square hole was made at the head of the grave, conveniently determined by the position of surrounding grave stones. Since the soil was loose and the box containing the coffin was usually only about four feet down, this task was easily accomplished.

Body snatchers rarely dug up the entire coffin. "No self-respecting grave robber," said Suzanne Shultz, author of *Body Snatching: The*

Robbing of Graves for the Education of Physicians in Early Nineteenth Century America, "would have loitered in a cemetery for the length of time it would have taken to accomplish this task." When the digger heard his shovel clunk on the wooden coffin, he would slip a crowbar under the lid and pry it upward until it broke, usually around chest-high. Sometimes the digger used an auger to bore holes in the lid to speed breakage. Axes, hatchets, or saws were to be avoided; they made too much noise.

To remove the body, the men would sling a rope around the neck or shoulders and pull the body through the opening in the lid, sliding it out of the coffin. Some resurrectionists preferred to use an iron bar with a crossbar handle and a hook at the other end that could fit under the jaw. Hooks often damaged the mouth, though, so care had to be taken to avoid injury. Clients preferred their cadavers unblemished by workmen's tools.

The men would next remove the body's clothes and then put them back into the grave, so as not to be charged with theft. "Neither the common nor the statutory law of England which governed the American colonies provided any penalty for exhumation of a human body," said Frederick Waite, professor emeritus of embryology and histology at Western Reserve University. "However, the taking of the shroud or other apparel constituted a felony under the common law, 'for the property thereof remains in the executor or whoever was in charge of the funeral.'" The body might not have had any inherent rights, but the clothing certainly did.

The body would then be placed into a sack or on a sheet of some kind, which often meant folding the body in half to fit, and occasionally in thirds. The grave would be refilled as quietly and as carefully as possible and the items on top precisely placed.

Citizens tried a variety of preventive measures to frustrate resurrectionists on their appointed rounds. If a family was wealthy enough, they might have their loved one placed into a vault or hire guards to watch over the grave until the body decomposed beyond the time an anatomist would want it. Families unable to afford either of those options might place boulders, heavy slabs of wood, or a plank of iron bars over the grave as a deterrent. The poorest families arranged stones, beads, buttons, or other items around the grave, so that if body snatchers disturbed the dirt the family would know. Body snatchers were clever, though, and usually found ways around those preventive measures. Boulders could be moved. If a stone was too heavy to move, body snatchers could dig down at the foot of the grave and remove the body through the small end of the coffin. Guards could be bribed or distracted, and markers could be replaced.

With the site restored, the men would load the body onto a wagon and transport it to a waiting anatomist, who paid the men, often handsomely. In the early 1800s, resurrectionists typically received somewhere between five and twenty-five dollars per body, a princely sum at the time. The price varied by the condition of the body, with the freshest bodies fetching more than partially decomposed ones.

Speed was critical for such adventures. Spending too long to remove a body increased the odds of capture by the local constabulary. James Blake Bailey, librarian of the Royal College of Surgeons of England in the 1890s, put together a "fragmentary" account of the "doings of one gang of the resurrection-men in London" over one year, 1811–1812. In his iconic book, *The Diary of a Resurrectionist*, he wrote, "The amount of time required for the work depended greatly on the soil. One man told Bransby Cooper that he had taken two bodies from separate graves of considerable depth, and had restored the coffins and the earth to their former positions in an hour and a half. Another man said that he had completed the exhumation of a body in a quarter of an hour; but in this instance the grave was extremely shallow, and the earth loose and without stones. If much gravel had to be dug through, the resurrection-men had a peculiar way of using their spades, so that the gravel was thrown out of the grave quite noiselessly."

Body snatching usually took place in the evening during winter. Scottish physician Sir Robert Christison explained his preference: "The time chosen in the dark winter nights was, for the town churchyards, from six to eight o'clock, at which latter hour the churchyard watch was set and the city police also commenced their night rounds. . . . Operations in the country were necessarily conducted at a later hour. Certain country churchyards were selected for convenience of approach and their distance from houses." Hot summer nights were the worst, not just because warmth speeds the process of decomposition but also because summer nights aren't as dark as winter nights. Even when a body was stolen during winter under ideal conditions, the speed of decomposition compelled that dissection be carried out as swiftly as possible.

There may be no field of study more interested in the process of decomposition than the forensic sciences. Until the early 1980s, pathologists and other forensic scientists lacked a consistent and accurate method of determining the length of time since death. That period, known as the postmortem interval, or PMI, can prove critical in solving murders. Famed forensic anthropologist William Bass became acutely aware of the problems with inaccurate PMIs during a murder investigation in 1977. The grave of a Civil War colonel named William H. Shy had been disturbed behind a home in Franklin, Tennessee. When police examined the grave, they found a hole in an old cast-iron coffin and an unusually well-preserved cadaver missing its head, a hand, and a foot. It seemed that Shy's body had been removed and another buried in its place. Bass was called to the scene to aid in the investigation. He had been studying how humans decompose in various environments since he first arrived at the University of Tennessee in 1971.

Bass determined that the individual in the grave was a white male about five feet, eleven inches tall, weighing 175 pounds. He was somewhere between twenty-six and twenty-nine years old and had died from a blow to the head. Bass also determined that the man had died six to eight months previous, which prompted police to begin a murder investigation. However, when Bass brought the body to his lab at the University of Tennessee, he discovered why the body had been so well preserved. It had been embalmed, which Bass had rarely encountered in his previous studies of human decomposition. "I got the age, sex, race, height, and weight right," admitted Bass, "but I was off on the time of death by 113 years." The body was indeed that of Col. Shy, who had died of a gunshot wound to the head. Apparently some body snatchers had broken into the coffin, pulled Shy's body through the hole, and stolen whatever they found of value. Then they covered the grave again and left. No arrests were ever made in the case, but Bass's experience had left him more determined than ever to learn as much as he could about human decomposition.

He established in 1987 the world's first facility to research the effects of trauma, water, weather, and various burial practices on decomposition. The University of Tennessee's Forensic Anthropology Center in Knoxville includes a two-acre area described by author Mary Roach in her fascinating and nimbly written book, *Stiff: The Curious Lives of Human Cadavers,* as "a lovely, forested grove with squirrels leaping in the branches of hickory trees and birds calling and patches of green grass where people lie on their backs in the sun, or sometimes the shade, depending on where the researchers put them." The grove is home to the Anthropology Research Facility (ARF), sometimes called a "body farm." The ARF holds 150 to 300 donated human cadavers at a time. The cadavers, lovingly referred to as donors, are left to decay in a variety of natural and controlled environments, allowing researchers to study the processes and time-table of decomposition.

Some of the cadavers donated to the center are placed fully clothed, some partially clothed, and some are naked. Researchers have "buried bodies in shallow graves," said Roach, "encased them in concrete, left them in car trunks and manmade ponds, and wrapped them in plastic bags. Pretty much anything a killer might do to dispose of a dead body the researchers at UT have done also." Those bodies are then studied closely by forensic scientists, medical examiners, and law enforcement officers to determine how the process of decomposition changes depending on each individual environment.

Decomposition begins within moments of a person's death when the heart, lungs, and brain cease functioning. The skin cools to the touch as postmortem cooling, or algor mortis, begins. Under normal conditions, the body temperature drops about 1.5 degrees Fahrenheit each hour until it reaches the temperature of the surrounding environment. This cooling continues for about six hours, depending on the body's temperature at death and the surrounding environment. The rate of cooling increases if the person was naked, thin, or immersed in water and decreases in someone thickly clothed or obese.

Along with that cooling comes a buildup of lactic acid in all body muscles, causing stiffening. This phase, called rigor mortis, appears from one to four hours after death and is apparent first in the muscles of the eyelids and jaw. All muscles, including those in the heart and intestines, become fully stiffened roughly twelve hours after death. Rigor mortis lasts for about twelve hours or so, depending on the environment, with cooler environments slowing the process.

Postmortem lividity, also called postmortem hypostasis, refers to the way blood pools in the body when it is no longer being pushed around by the heart. Blood in a person lying in bed at death tends to pool in the back, buttocks, and the back of the arms and legs, forced by gravity to those dependent areas. Blood pooling causes a distinctive discoloration of the skin over those areas, a discoloration that becomes apparent about an hour after death.

Throughout those processes, the body is breaking itself down. During life, cells release special enzymes that break down waste in the cells but which, after death, start breaking down the cells themselves, a process called autolysis. During autolysis, a process that lasts just a few days, cells throughout the body begin breaking down proteins and carbohydrates. Eventually the cell membrane breaks down, releasing the cell's contents. At that point, the process of putrefaction begins.

During putrefaction, bacteria, fungi, and other kinds of microscopic agents normally present in the body begin feeding off decomposed cells and multiplying aggressively. Around ten thousand species of bacteria live in each healthy adult, many of them residing in the gastrointestinal tract. Lita Proctor, a National Institutes of Health scientist and director of the ten-year-long Human Microbiome Project, explained that "humans don't have all the enzymes we need to digest our own diet. Microbes in the gut break down many of the proteins, lipids, and carbohydrates in our diet into nutrients that we can then absorb. Moreover, the microbes produce beneficial compounds, like vitamins and anti-inflammatories that our genome cannot produce."

In death, bacteria move from the gut to the blood vessels and travel throughout the body. Then the bacteria that can survive best without oxygen take over, and *Clostridium perfringens* reside firmly at the head of that pack. All of the more than a hundred species of *Clostridium* are anaerobic, meaning they can survive only in environments without oxygen. Those bacteria break down the tissues and organs of the body and cause visible and, for the squeamish, disturbing changes. Skin changes can prove particularly gruesome. "The liquid from the cells gets between the layers of skin and loosens them," explained world-renowned expert in human decomposition Arpad Vass. "As that progresses, you see skin sloughage. As the process progresses, you see giant sheets of skin peeling off the body."

The face puffs up as well, and the abdomen bloats from a buildup of gases released by the bacteria. Those gases—mostly hydrogen, methane, and carbon dioxide—are produced during life as well, but the body releases them through belching and flatulence, processes controlled by muscles. When the muscles stop working, the gases build up. Gases also build up in other parts of the body, particularly the lips, tongue, and breasts in females and penis and testicles in males. At some point, typically several days after death, the gases build up so much that the intestines give way and release their gases and liquids into the abdominal cavity. Occasionally the abdominal wall itself gives way, leading to what Vass described as a "rending, ripping noise." The bloating phase lasts about a week.

Some organs decompose more rapidly than others. The lungs and organs of digestion tend to decay early, along with the brain. The prostate and non-pregnant uterus decay later. As organs decay, they lose their shape and change in consistency. After about three weeks, individual organs within the abdomen and chest cavities may still be identified. "After that," said Vass, "it becomes like a soup in there." An acrid, rank-smelling soup.

The odors given off during putrefaction each have their own profile, with hydrogen sulfide's odor being the most widely known; it smells like rotten eggs. Other chemicals given off during decay include indole, skatole, cadaverine, and putrescine. All four are produced by the breakdown of amino acids, and all are nasty. Indole can smell like mothballs or, to some people, a wet dog. Skatole smells like feces. Cadaverine's odor has been described as akin to bad breath, urine, and semen, with putrescine's odor a pungent fishy odor or a biting ammonia. And all four of those fetid chemicals are found naturally in human saliva and are responsible for that well-known kiss killer, bad breath.

For people unfortunate enough to expire and not be found for a day or two, especially those who die outside, insects can enter the fray. Flies that land on the body can lay eggs, often near the mouth and nose. Within twenty-four hours, the eggs develop into larvae, or maggots, and begin feeding off the decaying flesh, hastening the process even more.

The rate of decay doubles for every increase of 50°F in temperature, so putrefaction during the summer months can be observed within thirty-six hours, sometimes less. The process occurs even faster in humid weather. The rate of putrefaction also depends on where the body is located. Forensic scientists often cite the work of Johann Ludwig Casper, the chief medical examiner of Berlin in the mid-1800s, who made detailed illustrations of bodies in various stages of decomposition. Casper developed what is now known as Casper's dictum, which states that a body left open to air decomposes twice as fast as a body submerged in water and eight times as fast as a body buried in the earth.

Overall, the process of putrefaction typically starts about eighteen to thirty-six hours after death, so the longer it takes resurrection men to grab a body, cover up their theft, and cart the body to an anatomist, the worse shape the body will be in. Generally, bodies older than ten days were worthless. Students in the 1700s and beyond would have been accustomed to cadavers in various stages of decomposition, but without refrigeration, learning anatomy from the bodies would have proved increasingly difficult as time marched on. Even so, bodies would be opened, cut up, and examined for weeks at a time, with students first learning superficial anatomy and then delving deeper into the body and its extremities.

Exactly how Richard Bayley and his colleague, Wright Post, conducted their anatomy classes remains unknown, but they probably conducted classes similarly to other anatomists of the time. Students would have applied to take an anatomy class, been accepted, and then been advised to obtain several instruments to perform dissections. "For example," wrote physician John Shaw in his book *A Manual for the Student of Anatomy*, "he could not dissect the nerves of the spine, nor of the head, without a small saw, two or three chisels of different sizes, a small mallet, and the strong pincers (that are used to pull out nails). . . . For the more minute dissections, he will require two small hooks and a sharp steel point. The etching tools which are used by engravers are very useful, particularly if the points are bent a little, as we can then easily tear away the cellular membrane from the small nerves."

Before students could view a dissection or take part in one, however, the body had to be prepared. If the instructor didn't have enough cadavers, he would sometimes have the body cut into pieces so one or two students could be working on, say, a leg, while other students could be working on an arm or a head. This was especially true in Britain in the early 1800s, when there were a great many students and nowhere near enough cadavers to study. "They might have a knee one day, an elbow another," said Jenna Dittmar, a biological anthropologist at Edward Via College of Osteopathic Medicine in Monroe, Louisiana. "As long as they dissected two complete bodies they could be medically licensed. It didn't matter if it was in bits."

Franklin Dexter, a Harvard anatomist, provided his students with cotton cloth to ward off decomposition. One student wrote in his notes for anatomy class, "Cotton cloth (10¢ for 2 yds) given out at 1.30. Keep moist & wrap part in it. All parts given out at 2 P.M.— if nothing done in 24 hrs or demonstrator notified, parts forfeited. Tickets stamped 'Bond' must be shown or parts forfeited."

Dissecting body parts properly requires students to identify the various components of the body, such as veins, arteries, nerves, and so forth. In a living person, arteries appear pale orange, round, and full, pulsating with blood. Veins appear more reddish, like an uncooked chicken breast, and nerves are white, firm, and appear stringy. In a cadaver, though, the blood vessels have collapsed and are not pulsing. They also take on a kind of muddy color, along with nerves, making identification tricky. Injecting dyed fluids into the blood vessels would make identification far easier. Instructor John Shaw used a particular formula for making a red fluid for large arteries: fifteen ounces of yellow beeswax, eight ounces of white resin, six ounces of turpentine varnish, and three ounces of vermillion.

The beeswax, resin, and turpentine would be heated and mixed with a wooden pestle. Vermillion mixed in a different pot would then be combined, over heat, with the liquid ingredients. The preparer would use care to avoid overheating the mixture, which could burn the arteries when injected. After the mixture reached the proper temperature, the preparer would fill a syringe with the fluid and inject it into the appropriate artery. The procedure would be repeated for veins, using King's yellow, Shaw's preferred color. The mixtures would slowly solidify, providing the vessels with color and bulk.

Sometimes injections would be completed during preparation of the body part; other times during the dissection itself. "The arteries

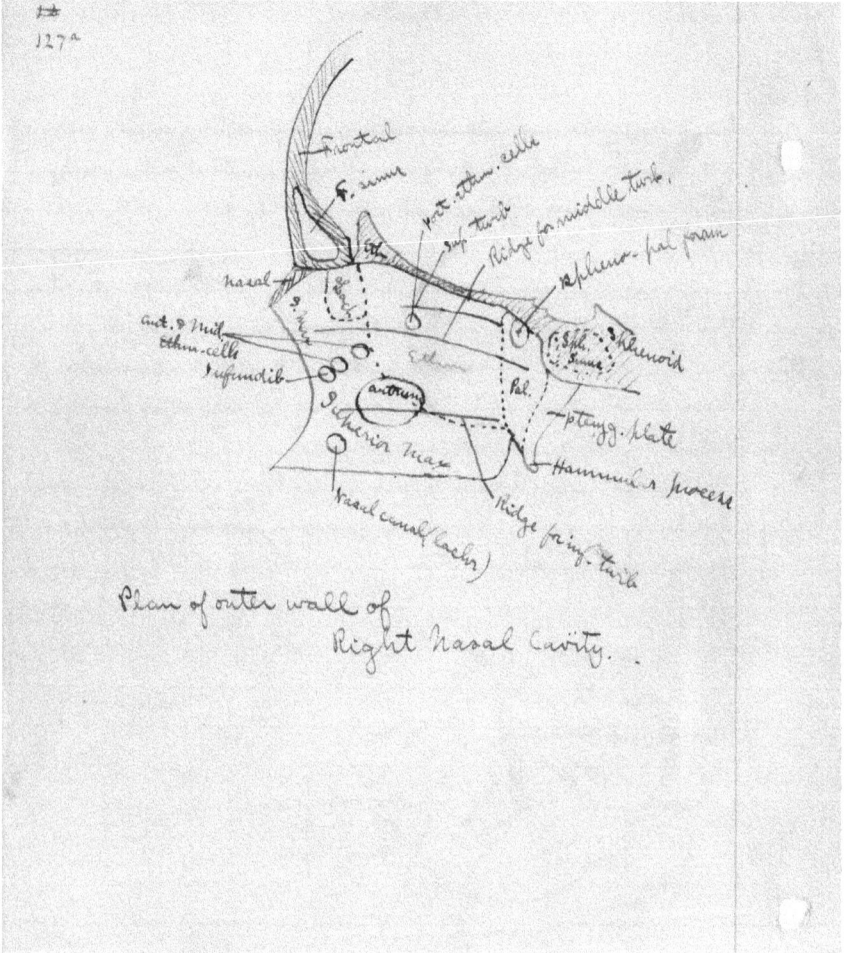

Drawing by Harvard medical student Ralph C. Larrabee of the outer wall of the right nasal cavity, drawn in 1893. Larrabee graduated medical school in 1893 and practiced in Boston. He was an avid mountaineer and longtime member of the Appalachian Mountain Club. COURTESY OF HARVARD MEDICAL LIBRARY.

may be injected after the muscles of the abdomen have been dissected," wrote Shaw in describing the steps in the dissection of the abdominal cavity. "To do this neatly, we should tie the aorta above the diaphragm, and also one of the iliac arteries at its origin from the aorta, and then put a pipe into the other common iliac, as close to the aorta as possible, so that there may be enough of the artery left to enable us to put a tube into it afterwards for the injection of the lower extremity." Shaw recommended a different injection mixture for abdominal vessels, preferring "a strong solution of glue, coloured with red lead, or an injection made of tallow and turpentine varnish. As both of these compositions must be used while warm, it is necessary to heat the vessels of the abdomen. This is most easily done by making an opening into the intestines and injecting a quantity of hot water into them."

When students and professors had finished working with a cadaver, few options existed to dispose of the remains. Sometimes anatomists saved body parts for their collections, storing them in jars filled with oil of turpentine or some other liquid as a preservative. (Formaldehyde wouldn't be invented until 1859.) Sometimes the flesh would be boiled down so that the bones could be dried and preserved as a skeleton. Most of the time, though, the remains of dissected bodies were either burned or buried somewhere, often with multiple body parts buried in the same receptacle. Infants were often placed into coffins with another body or a collection of body parts, not always from the same person.

Descriptions of Bayley's anatomy lab have not survived, but it likely would have been similar to Clossy's lecture room at King's College. Clossy taught in a room on the third floor; Bayley's was on the first, possibly the second, floor. Clossy kept an extensive collection of anatomical specimens in an adjoining room. Bayley had also brought specimens home from his time in London, though the number is unknown. He was given permission to use just one or two rooms in the otherwise empty New York Hospital, so he kept his supplies in cabinets in the lab or perhaps in a nearby room. Both physicians undoubtedly kept their instruments, chemicals, dyes, cloth, and other pertinent items at the ready, so students laboring at cadavers on tables could grab supplies as needed.

A British artist named Thomas Rowlandson painted his view of a dissecting room in 1815 or 1816, called "Death in the Dissection Room." Rowlandson's depiction of a working anatomy laboratory reflects the popular view of physicians as ghoulish and disrespect-ful of death. The doctor and his busy staff are conducting at least two dissections, one in the foreground and one in the background, while more bodies await preparation. One female body is callously left lying on the floor, and another, still in a bag, arrives through the door. The delivery man looks anxiously behind him, suggesting that the new body has been stolen or that the laboratory is operating outside of the law. Skeletons and specimen jars fill the room, and entrails and instruments (including the satirically ubiquitous clys-ter syringe) clutter the floor. A human skeleton representing Death ambushes the doctor with an arrow, suggesting that the body next in line for the dissection table will be his.

Neither Clossy nor Bayley would have seen Rowlandson's paint-ing by 1788, but they were certainly aware of the stigma associated with human dissection and the anger that could target anatomists. They continued their teachings regardless, because they knew that anatomy was then, and remains today, a critical course for all medical

Satirical illustration of Death interrupting a surgeon and his staff performing two dissections, with a third (woman on floor) lying nearby. Painted by Thomas Row-landson. COURTESY OF THE NEW YORK PUBLIC LIBRARY.

students. The dissections performed in an anatomy lab allow students to make incisions, trim one layer of tissue away from another, isolate organs and blood vessels, and visualize the various ways diseases could manifest in the body. Clossy published a text in 1763 called *Observations on Some of the Diseases of the Parts of the Human Body: Chiefly Taken from the Dissections of Morbid Bodies.* In it he presented a series of patients who had experienced some disease or injury and then died. He then explained how the person's anatomy had changed as a result of the disease.

In one case, a man about age thirty had "received a stroke with a pole on the left parietal bone, which fractured the skull to splinters, and wounded the membrane [meningeal membrane, a three-layered membrane covering the brain and spinal column]." Despite treatment the patient died seventeen days after injury. Clossy then described the anatomy of the injured area for his readers: "Now at one place under the . . . fracture, the pia mater [delicate innermost layer] appeared black, and on slitting the membrane a clot of blood was discovered, which had formed for itself a cavity in the substance of the brain, as large as a pullet's egg. Blood, therefore . . . may not only diffuse itself so far as to fall between the hemispheres, but [also] to force its way into the substance of the brain, and there accumulate."

Bayley might well have had similar fracture cases, and if so, his students would have been able to explore the fracture and see and feel whatever tissue damage had occurred. In such cases, the students were at the same time learning pathology, the study of disease, right along with anatomy. They might also have been able to view any number of illustrations of cranial anatomy, fractures, lacerations, and contusions, and whatever specimens from previous dissections might have been stored in the lab. In that way, the lab was the central focus of education, and without it, their learning would have slowed greatly, if not stopped altogether.

The rioters who descended on Bayley's lab in April 1788, however, were furious and cared not at all for anyone's education. They wanted revenge.

CHAPTER 8
BUILDING A NATION

THE YEARS PRECEDING 1788 WERE REPLETE WITH DRASTIC changes in how Americans lived, especially in New York City, which had been a British stronghold since the earliest days of the war. Not long after the Treaty of Paris was signed on April 15, 1783, the commander-in-chief of British forces in North America, Sir Guy Carleton, received orders to begin the evacuation of troops from occupied areas. He was determined that all troops would be gone by November 22 that year, and he almost made it; the last British troops didn't leave the city until November 25. Many citizens who had fled the city early in the war, such as brewery owner Elizabeth Rutgers, returned to New York to begin anew. The destruction they found must have stunned them.

A raging conflagration in 1776 had demolished hundreds of homes and buildings. The fire occurred just days after the British army took control of the city at the start of the war. Possibly started by Patriot sympathizers, the fire began at the southern tip of Manhattan, moved diagonally through the center of the island, and then fanned north along the Hudson River. It destroyed Trinity Church, leaving nothing but a shell. Another enormous fire three years later destroyed hundreds more buildings, this time along the city's east side. By war's end, more than a third of the city lay in ruins. William Alexander Duer, a city attorney, recalled that the burnt-out wreckage of buildings "cast their grim shadows upon the pavement, imparting an unearthly aspect to the street." James Duane, who would become mayor in 1784, returned in November and found two of his previously occupied homes in horrible shape, "as if they had been inhabited by savages or wild beasts."

Into that chaotic, battered city rode the commander-in-chief of the Continental Army, George Washington, along with about eight hundred of his troops, to reclaim the city. He stayed only a brief time, telling his generals before he left on December 4, "With a heart full of love and gratitude I now take leave of you. I most devoutly wish that your latter days may be as prosperous and happy as your former

A lithograph by Christian Inger depicts George Washington riding triumphantly into New York City on horseback on November 25, 1783. The war had ended in 1781, but the Treaty of Paris had been two years in the making. Shown in this famous image are Washington's principal officers, on horseback far left are Gen. Nathaniel Greene, Gen. Israel Putnam, Gen. Henry Knox, Alexander Hamilton, Gen. Marquis de Lafayette, Gen. Thaddeus Kosciusko (with flag), Gen. Horatio Gates (directly under tip of flag), George Washington, Baron Friedrich Wilhelm von Steuben, and Gov. George Clinton. Seated left to right on small balcony at left are Abigail Adams, Polly Caton (young socialite from Baltimore), and Miss J. Marshall. On the balcony at right are, from left, Miss Bingham (daughter of Philadelphia socialite Anne Willing Bingham), Martha Washington, Cornelia Clinton, Anne Willing Bingham, and Elizabeth Hamilton. Sarah Jay is seated at lower right, with hat.
IMAGE COURTESY OF LIBRARY COMPANY OF PHILADELPHIA.

ones have been glorious and honorable." Congress took up residence during the following spring, first in City Hall and, later, with usable space at a premium, branching out to other public buildings and, in one case, to a pub called Fraunces Tavern at 54 Pearl Street.

Over the ensuing five years, the city underwent explosive growth. The cornerstone for St. Peter's Church, the city's first Catholic church, was laid in October 1785 and proved a bellwether event for a population of Catholics that had roughly doubled since the end of the war. Merchants from New England poured into the city and set up shop. Others came from throughout Europe and the British Isles, intent on tapping into New York's promising marketplace. Among them was a young German merchant named John Jacob Astor, who arrived in March 1784 and would soon begin selling imported musical instruments from his shop on Queen Street. Eventually he entered the fur industry and rose to become one of the wealthiest men in early America.

Destroyed buildings were rebuilt, land was bought and sold, and new buildings were erected to accommodate a growing population and the businesses it supported. The city served as the temporary capital and, over time, became the center of trade and commerce. The first bank in the city was formed in 1784, with Hamilton alone drafting its constitution, a feat Hamilton biographer Ron Chernow called "herculean." The constitution served as a template for many banks thereafter, and the Bank of New York became a key player in rebuilding the city's damaged economy.

Slavery proved a relentlessly thorny issue. Hamilton, Governor Clinton, Mayor Duane, John Jay, and other leaders who had spoken out against the practice met in early 1784 to create the New York Manumission Society, with Jay being elected president. Jay owned five slaves, and Clinton owned eight. In fact, half of the new society's members owned slaves. Jay had been an outspoken critic of slavery for many years and would fight the despicable practice for the rest of his life. Yet he owned humans, as did his father before him, and believed in an incremental abolition of slavery. The practice held such a significant position in the nation's economy that many abolitionist slaveholders couldn't see a way to rid the country of it without destroying the economy. Thomas Jefferson might have summed up the conundrum best when he said, "We have the wolf by the ear, and we can neither hold him, nor safely let him go. Justice is in one scale, and self-preservation in the other."

The issue of slavery would have to wait; the nation didn't even have a constitution yet and wouldn't until the document was fully ratified on June 21, 1788. In the meantime, the nation's war debt was ballooning to alarming levels, and anger among the citizenry was smoldering. Farmers were especially affected. Asked to increase crop production for the war, many farmers obtained loans to expand their farms. The food they produced fed three armies and three navies—those of the British, French, and Americans. After the war, the need for food dropped precipitously, and much of the food raised sat unsold and spoiled. Unable to sell their wares, many farmers suffered. "They were in deep financial difficulty," said Constitutional scholar John Kaminski. "Much of their land was being seized and sold at public auction to pay for their debts and to pay taxes. The farmers asked, 'What did we fight the American Revolution for? Life, liberty, and property. We can't pay our debts. We're being arrested. Our land is being taken away, and we're put into debtors' prison if the sale of the property will not bring in enough money to pay the debts and the taxes.'"

All of that anger erupted in 1786 when a former Army captain named Daniel Shays led a six-month insurrection in western Massachusetts. Many farmers in the state held great debts left over from the war, and the legislature in Massachusetts had been unresponsive to calls to forgive those debts and to print more money to boost local economies. With an increasing number of farmers being arrested and imprisoned over their debts, Shays and his army—soldiers-turned-farmers, some with guns, some with pitchforks—took over the Court of Common Pleas in Northampton, Massachusetts, on

August 29. That action propelled other angry citizens to take over courthouses around the state.

In need of weapons and ammunition, Shays and his men headed to the state's arsenal in Springfield. A group of soldiers under the command of Benjamin Lincoln, a general in the Revolutionary War, also traveled to Springfield to defend the arsenal. The troops were hired through a fund of $20,000, money raised by Lincoln and donated by wealthy Bostonians and the elite of surrounding towns. The Springfield arsenal held 7,000 small arms with bayonets, 1,300 barrels of powder, and plentiful ammunition. Lincoln reached the arsenal first. On January 25, 1787, Shays' men attacked. A brief battle ensued. Four of Shays' men were killed, a large number wounded, and the remainder scattered. The uprising was over, though the anger seething underneath remained.

The rebellion sparked fears in General Washington and other leaders that the Articles of Confederation were simply too weak and ineffectual for the new nation. Congress authorized a convention in Philadelphia to examine the Articles and "amend" them only. The resultant Constitutional Convention in the summer of 1787 would discard the Articles completely and instead write a new constitution for the nation.

Washington, reluctant at first to attend, finally relented, in part to prevent other Shays Rebellions from happening. He was named the convention's president and officiated over a four-month-long meeting among many of the nation's most preeminent leaders, including James Madison and his great friend Alexander Hamilton, the aged, brilliant Benjamin Franklin, Virginia's George Mason, and the "Penman of the Constitution" Gouverneur Morris, the "whimsically flamboyant" delegate for Pennsylvania. Seventy-four delegates had been invited to the convention, and fifty-five attended.

Rhode Island refused to send delegates, fearing that the convention was a ruse to overthrow the government. Patrick Henry, perhaps the nation's finest orator, refused to go, saying that he "smelt a rat in Philadelphia." The rat he feared was that the convention would lead to an overly powerful centralized government and feeble state governments. Plans at the convention were proposed, debated, and rejected. Deadlocks abounded. Washington, at one point, decried the "narrow-minded politicians" and their inability to see the grand picture. Three aspects of the draft had proved particularly mettlesome to the delegates. First, the document called for a strong central government with three branches: the legislative, judicial, and executive. It gave the federal government enormous power and state governments considerably less. That alone was seen by some as being too similar to the recently overthrown centralized British government. They believed that an alliance of strong states, rather than a centralized government, would provide the best protection for the nation's liberty.

Second, that most detestable of practices, slavery, remained impossible to solve. Northern states wanted to exclude the enslaved from being measured in state populations if they weren't also going to be granted rights. Said Hamilton during the negotiations, "They [slaves] are persons known to the municipal laws of the states which they inhabit, as well as to the laws of nature. But representation and taxation go together—and one uniform rule ought to apply to both. Would it be just to compute these slaves in the assessment of taxes; and discard them from the estimate in the apportionment of representatives? Would it be just to impose a singular burthen [burden], without conferring some adequate advantage?"

The less-populated Southern states, however, could see no way to survive without slavery. They produced too many goods, too valuable for trading to eliminate the practice. Pierce Butler, a delegate from South Carolina, countered that "the labour of a slave in S. Carola. was as productive & valuable as that of a freeman in Massts." He added that "an equal representation ought to be allowed for them in a Government which was instituted principally for the protection of property." Southern delegates also wanted more representatives for each state, so they pushed to include slaves in the population. The delegation finally agreed to a compromise, the so-called three-fifths compromise, which stated that the House of Representatives would be divided according to the number of free people in each state, plus three-fifths of the number of enslaved. The compromise enraged abolitionists and would eventually be overturned, but not for more than eighty years.

Third, the document omitted a bill of rights, considered an egregious error by Mason and Jefferson. "A bill of rights is what the people are entitled to against every government on earth," Jefferson wrote to Uriah Forrest, a Maryland delegate, "and what no just government should refuse." Those who supported the document, principally Madison, argued that because the government can exert only those powers specified in the Constitution, a separate bill of rights wasn't needed. Madison would later reverse course and drive the creation and signing of the Bill of Rights in 1789.

The eminent Dr. Franklin gave the Constitution his approval, though reluctantly. He said there are "several parts of this Constitution which I do not at present approve." He added, however, that "the older I grow the more apt I am to doubt my own judgment and pay more respect to the judgment of others." He then urged ratification, which required signing by nine of the existing thirteen states.

Gouverneur Morris, a talented orator and graduate of King's College, polished the draft, eloquently so. He condensed twenty-three separate articles down to just seven and added perhaps the most elegant, profoundly written sentence in American history, the Preamble. "We the people of the United States," penned Morris, "in order to form a more perfect union, establish justice, insure domestic tranquility, provide for the common defense, promote the general welfare, and secure the blessings of liberty to ourselves and our posterity, do ordain and establish this Constitution for the United States of America." The Constitution contained so many new concepts, proposed such an unexpected form of government, and addressed so many issues that had bedeviled society as a whole that ratification was sure to prove anything but guaranteed.

The Constitution was first printed in New York on September 21, 1787, four days after the convention ended, by the *New York Daily Advertiser*, the nation's third newspaper. The paper was published by Francis Childs, one of Franklin's former partners in Philadelphia. When the Constitution reached the populace, it immediately became a topic of conversation throughout the city. Discussions in taverns and pubs exploded into heated arguments. Many newspaper essays, the social media of its day, criticized the document and its creators, sometimes scathingly so. Other essays supported the document in robustly glowing terms, a constitution brought back from the Mount.

Many of the editorials were signed with a pen name, a common practice. Essays signed "Cato" were almost certainly written by Governor George Clinton, who often took aim squarely at Hamilton, and Hamilton, the assumed writer "Caesar," responded in kind. The two men had been bitter political enemies since the early 1780s, and the convention did nothing to change that. Hamilton was a federalist, one who believed in a powerful centralized government; Clinton was solidly against the idea and spearheaded the antifederalist movement in the city. Essays by Cato and Caesar, as well as their respective supporters, began appearing in the *Daily Advertiser*, *New York Journal*, *Northern Sentinel*, and other newspapers during the convention and continued for weeks afterward as the citizenry grappled with the new national charter.

Federalists pointed out that the document would "render us safe and happy at home, and respected abroad." They admitted that even though the document wasn't perfect, it was "much more so than the most friendly and sanguine expected." Antifederalists argued that the Constitution was "a monster with open mouth and monstrous teeth ready to devour all before it." As the document spread, society began to split into federalists and antifederalists, each faction presenting apocalyptic visions should its side not win. "The federalists evoked disunion, civil war, and foreign intrigue," said Chernow, "along with flagrant repudiation of debt and assaults on property.

The antifederalists talked darkly of despotism and a monarchy, the ascendancy of the rich, and the outright abolition of the states."

Hamilton grew increasingly worried that his home state wouldn't ratify the document and that Clinton and his antifederalists would prevail, an outcome he could not abide. He decided to write a series of essays that would lay out all of the reasons why the Constitution should be accepted and respond to each argument against it. He enlisted Jay, Madison, and two others, whose contributions would not be used, to join him in this prodigious endeavor. Hamilton wrote the first essay in the series, which appeared October 27 in *The Independent Journal.* Titled "The Federalist 1" and signed Hamilton's alias, "Publius," the essay introduced the series, which would turn out to consist of eighty-five entries. In the introduction, Publius urged citizens not to be swayed for or against the charter by any opinions "other than those which may result from the evidence of truth." He wrote that his "arguments will be open to all, and may be judged of by all," and then presented the topics to be covered in subsequent essays.

The remaining essays, all published under the name "Publius," were written between October 1787 and May 1788. Jay wrote Federalists 2 through 4, dealing largely with foreign affairs, but then in November, he was forced on hiatus after an acute attack of rheumatism, today known as rheumatoid arthritis. Jay told Washington in a February 1788 letter that "a constant pain in my left side continues," adding that, regardless, he was "happy that my long and severe illness has left me with nothing more to complain of." Jay's rheumatism let up long enough for him to write one final essay, No. 64, "The Senate as a Court of Impeachment," published March 7, 1788, a little over a month before his heroic actions during the Doctors' Riot.

Hamilton and Madison churned out dozens of essays, some 175,000 words, with remarkable speed and enduring eloquence. With all essays signed Publius, historians had to determine which founder wrote which essay. As of this writing, Hamilton is said to have written a staggering fifty-one essays on his own, Madison fifteen, and

Jay five. Hamilton and Madison worked together on three papers, with the remaining eleven having been written by either Hamilton or Madison. Most of the essays were published in either the *New York Packet* or *Independent Journal* and reprinted in numerous other newspapers in New York and other states. The authors hoped that the Federalist Papers would propel New York to ratify the Constitution, which it finally did on July 26, 1788, though historians have debated how much the documents actually helped the cause.

Regardless, the words of Jay, Hamilton, and perhaps especially Madison have long endured as a basis for legislation and legal decisions. Many essays continue to be used today to support decisions in a wealth of matters. Ira C. Lupu, a nationally recognized scholar in constitutional law, said that Federalist No. 78, which urged that the judicial branch remain independent of the other two branches, "is probably better known among American law students, legal academics, judges, and lawyers than any other single paper."

Many sections of the other essays have endured as well, perhaps none more so than this sentence in Federalist No. 50 ("On Maintaining a Just Partition of Power Among the Necessary Departments"): "If men were angels, no government would be necessary." The author, either Hamilton or Madison, historians aren't sure which, then expanded on the thought: "If angels were to govern men, neither external nor internal controls on government would be necessary. In framing a government which is to be administered by men over men, the great difficulty lies in this: You must first enable the government to control the governed; and in the next place oblige it to control itself." Simple, powerful words to support a compelling document that would form the basis for the entire system of government for a brand-new nation.

The Constitution addressed a great many issues, but not slavery, which had been a persistent thorn in the backside of the Founders. The first enslaved Africans arrived in the colonies in 1565 on a ship landing in Saint Augustine, Florida. More than fifty years later, another ship, the *White Lotus*, would arrive in Jamestown, Virginia, and would depart after having traded the lives of fifteen women and seventeen men for food. New York City received its first enslaved Africans when it was still called New Amsterdam. Paulo Angola, Cleyn Anthony, Simon Congo, Jan Francisco, Jan Fort Orange, Manuel de Gerrit de Reus, Gracia, Groot (Big) Manuel, Cleyn (Little) Manuel, Anthony Portugis, and Peter Santome were first to arrive, in about 1626. Three enslaved African women arrived shortly thereafter and, together with the original eleven, formed the first African community in the town. Between 1776 and 1800, more than 1.9 million people in Central Africa were captured, enslaved, and brought to the Atlantic region of America. Most of them came from areas known today as Senegal, The Gambia, Ivory Coast, Benin, Togolese Republic, Nigeria, Gabon, Republic of the Congo, and Republic of Zaire.

The African community in New York was a tightly knit one that often came together to celebrate life, their friends and families, and the cultures and traditions they shared. They prayed together in their own churches, the first being the Wesley Chapel on John Street, which eventually became the John Street United Methodist Church. This "plain and decent" church was founded by Philip Embury, an immigrant from Limerick, Ireland. African residents soon began attending church and sat among their White brethren. At some point, two formerly enslaved Africans, James Varick and Peter Williams, became church officials. The appointment raised the hackles of numerous White members of the congregation, and soon Black churchgoers were forced to sit in the back.

Other churches with a predominantly White congregation forbade Blacks from attending their services. Members of one Dutch Reformed church in Brooklyn refused to allow Blacks to become members because they "have no souls" and "they are a species different from us—witness their nauseous sweat, complexion, and manners etc. I cannot endure them near me. I would be asham'd to commune with them." At the time, three out of four households in the area included at least one enslaved African.

Free and enslaved Africans survived in the city, but the majority of them didn't come close to thriving. Malnutrition among enslaved Africans in New York was considerably more common than the non-enslaved. Infectious diseases also riddled the enslaved community, particularly tuberculosis, flu, typhus, syphilis, meningitis, osteomyelitis (bone infection), pertussis (whooping cough), and diphtheria. Enslaved Africans also suffered traumatic injuries frequently. Bone fractures, joint dislocations, head injuries, and all manner of cuts, burns, and bruises were caused either in accidents or, more likely, at the hands of their enslavers. White citizens, even though exposed to the same infectious diseases and traumatic injuries, tended to live remarkably longer than enslaved Africans, with nearly four times as many Whites living past age fifty-five as enslaved Africans.

Between 1630 and 1795, some 15,000 to 20,000 Blacks were buried in a six-acre area at the top of a hill just north of a large open area, called the Common, and east of Broadway. The city used the Common for activities that the more elite in New York wanted held away from them, a colonial form of the "not in my backyard" way of thinking. The Common held a tannery, brewery, slaughterhouse, prison, almshouse, and burial site for Blacks, the poor, prisoners, and deserters—the "unwanted, the smelly, and the unsightly" of the city. Eighteenth-century historian David Valentine described the area chosen for the cemetery as a "desolate, unappropriated spot, descending with a gentle declivity towards a ravine which led to the Kalkhook [in English, Collect or Fresh Water] pond."

Most of the area, labeled on a 1755 map as "Negros Buriel Ground," today lies beneath buildings and roadways. However, a small section of the Negroes Burial Ground was discovered in 1991 during construction and subsequently studied by a team of archeologists and other scientists from Howard University. Today, the African Burial Ground National Monument stands near the original burial ground and provides visitors and scholars with information

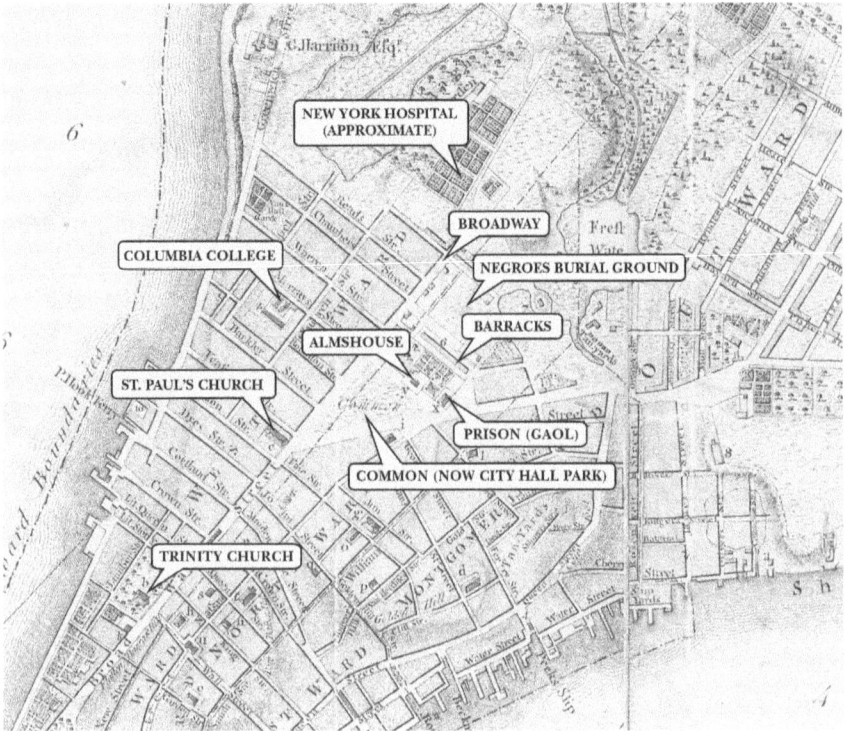

A section of a map by Bernard Ratzer in 1776 showing locations of key structures in the area where the Doctors' Riot occurred. Ratzer was a British cartographer sent to survey the East Coast during the French and Indian War. His map was used extensively for the British during the Battle of Brooklyn and later for capturing Manhattan. The most accurate of all city maps to that time, there are only four known copies in existence. COURTESY OF THE LIBRARY OF CONGRESS AND WITH GRACIOUS THANKS TO MARIE O'NEILL.

about slavery in New York City and some of the bodies buried on the grounds.

Most of the bodies, archeologists discovered, were buried in unmarked or crudely marked graves, with lines of small stones or a large rock placed at the head of the grave. Sometimes rings, pendants, and other pieces of personal jewelry were buried alongside the individual. Coffins, typically hexagonal in shape and made of inexpensive wood that decomposed quickly, were usually provided by the enslaver of the deceased and represented the enslaver's only contribution to the family at death.

In Western Africa, only kings were buried in coffins. Ordinary people were buried wrapped in a mat or shroud. Coffins became used more often for commoners throughout the African Diaspora during the eighteenth century. Even so, many male adults were buried without a coffin. In the area comprising the African Burial Ground, fully 75 percent of male bodies had been buried without a coffin. Historians believe those individuals were most likely Revolutionary War soldiers and refugees whose families or friends couldn't afford a coffin. Some of the bodies could also have been buried without a coffin because the owner refused to provide one. Many bodies were wrapped in cloth for burial, with the cloth held together by stitches or, for families with means, copper pins.

Nearly all bodies were buried, coffin or no, in a predominantly east–west alignment, with the head on the west side and the body supine, generally with arms at the side or crossed at the waist. A few individuals were buried with decorative items, such as beads, cufflinks, and rings, though most were not. Clam and oyster shells were sometimes included in burials, which "suggests the continuation, in at least some aspects, of African spirituality and burial customs."

Burials at the Negroes Burial Ground grew to become immensely important events for the African community in the city. "It was the place where kin were buried," reads a report on the burial site, "a place to memorialize and interact with the ancestors, a place to develop and strengthen neo-African identities, and a place to foment resistance. In essence, it was sacred space that was in many ways African to the core. Despite (or perhaps because of) its importance to people of African descent, the African Burial Ground was repeatedly desecrated by historical-period waste disposal and other regrettable activities." The burial ground might have been considered sacred by African families, but to the students of Dr. Richard Bayley, studying anatomy in a lab at an otherwise empty New York Hospital, it was just a cemetery where they could find recently buried bodies to dissect.

CHAPTER 9
TENSIONS SURGE

BODY SNATCHING HAD BEEN A PROBLEM IN NEW YORK SINCE AT least 1763, when the public became aware of Samuel Clossy having routinely dissected the bodies of enslaved Africans snatched from Black cemeteries. Alexander Hamilton, who had attended lectures by Clossy at King's College, might well have watched one or more of those dissections. Newspapers accused Clossy of grave robbing, yet he was still granted a permanent appointment at King's.

Clossy moved to London in 1780 and died there in 1786, never to work for Columbia College, the renamed King's. The medical school at King's had closed at the start of the war and then, as Columbia, appointed new medical faculty in 1784. However, no further work on the school was done until 1792, when Columbia reorganized the medical school and appointed nine professors under the super-vision of Samuel Bard. Between those times, anatomy was taught informally and unofficially by such physicians as Charles McKnight, Richard Bayley, and Wright Post. Nicholas Romayne taught the subject on his own. As a Columbia trustee, Romayne couldn't teach at the college, so he used rooms at the Almshouse and at Bridewell prison, located slightly east of Broadway in today's City Hall Park. McKnight was a surgeon who had trained in Philadelphia under William Shippen and served as a chief hospital physician during the war. Afterward he moved to New York and began giving lectures in the anatomical theater at Columbia. Bayley, who would later be granted a professorship at Columbia, gave his classes at New York Hospital. Like anatomy professors elsewhere, they all depended on cadavers from nearby cemeteries and hospitals.

The theft of bodies from Black cemeteries had never caused much concern for the White community, nor had the theft of bodies from potter's fields. One citizen, writing to the editor of the *Daily Advertiser*, claimed that "the only subjects procured for dissection are the productions of Africa, or their descendants, and those too who have no friends and have been transmitted to gaols, and elsewhere for having been guilty of burglary and other capital crimes. And if those characters are the only subjects of dissections, surely no person can object." Body snatching was easily rationalized by citing the benefits gained by anatomy students learning their craft. For instance, a newspaper in 1763 noted that after a Black man accused of rape had been killed, "the body has since been taken up and likely to become a raw head and bloody bones by our tribe of dissectors, for the better instruction of our young practitioners."

Physicians supported the practice as well, though in a backhanded way. Bard's father, John, also a physician, urged city leaders to pay closer attention to graveyard security, though not to prevent body snatching. Bard claimed on February 23, 1784, in a missive to a semi-weekly newspaper called *Loudon's New-York Packet*, that graveyards present "more danger of infection from a corpse than before death," and that "extreme activity of putrid effluvia" ignited dangerous fevers. Don't leave bodies in the ground, he believed, they're dangerous.

Other physicians seemed to urge that the dead be handed over to anatomists directly and that burial of such bodies be avoided completely, or at least within the city proper. A letter from "A Physician But No Practitioner" appeared in the *Packet* a month after Bard's letter and argued that toxic gases from cemeteries killed thousands of people and that the dead should be buried in areas well outside the city's center, not in church cemeteries. Another writer, "A Physician," followed up two months later, writing "It is certainly an act of indecency, if not a profanation, to leave the dead to rot in a place consecrated to divine worship." All three writers seemed to believe it was better to steal bodies and hand them over to physicians than to leave the bodies rotting in hallowed ground.

No religious leader of any denomination responded to the letters, owing at least in part to the still labile relationship between the church and the state. "New York had not yet ratified the Constitution," explained historian Robert Swan, "and the Church anxiously awaited the outcome of unsettling turmoil in separation of its affairs from the state." No longer governed by the Church of England, the Episcopal Church in particular tried to keep a low profile, not wanting to interfere in civil or state affairs. Most Christian churches were required to create burial grounds even before they built the church itself. Every church in the city had its own burial ground next to or behind the edifice. Trinity Church, with its front doors facing east to Broadway, created its churchyard on its north and south sides. Both White and Black church patrons were buried in Trinity's churchyard until an edict was issued on April 12, 1790, ordering that "in future no black Persons be permitted to be buried in Trinity Church Yard, nor any except Communicants in the Cemetery at St Paul's."

Trinity officials did, however, decide in 1773 that "a piece of the church lands should be let apart and appropriated for a burial ground for the Negro's." Swan explained that "interment in the black grave yard was tantamount to excommunication from the church without a trial. Thus, Trinity's families, more significantly, pious owners of devout, faithful slave church members, refused to permit their loved ones to be interred in unholy ground, a profanation of God's covenant." Church officials selected an area "bounded by Church Street, Reade Street, Chapple Street and the ground of Anthony Rutgers way," as the most appropriate for the site, which remained in use until city limits began moving north in the 1790s.

No matter where Black New Yorkers were buried, body snatchers followed. Resurrection men pilfered innumerable Black bodies from the Negroes Burial Ground. Angry and frustrated over resurrectionists stealing the bodies of their loved ones, Blacks in the city railed against body snatching as much and as openly as they could. On February 3, 1788, a petition arrived at the Common Council that had been signed by two thousand free Blacks and one thousand enslaved Blacks. The petition, deferential and carefully worded, urged the city to intervene.

> *Most humbly sirs, we declare that it has lately been the practice of a number of young gentlemen in this city who call themselves students of the physic to repair to the burying ground, assigned for the use of your petitioners. Under cover of the night, in the most wanton sallies of excess, they dig up bodies of our deceased friends and relatives of your petitioners, carrying them away without respect for age or sex, mangle their flesh out of a wanton curiosity, and then expose it to beasts and birds. Your petitioners are well aware of the necessity of physicians and surgeons consulting dead subjects for the benefit of mankind. Your petitioners do not presuppose it as an injury to the deceased and would not be adverse to dissection in particular circumstances; that is, if it is conducted with the decency and propriety which the solemnity of such occasion requires. Your petitioners do not wish to impede the work of these students of the physic but most humbly pray your honors to take our case into consideration and adopt such measures . . . to prevent similar abuses in the future.*

The petitioners' desperate plea might have been read at the council's next meeting, though no records exist to confirm it. Even if it was read, the mayor, James Duane, as well as all council members at the time, owned enslaved Africans, a situation that almost certainly colored their reception of the letter. The council either missed the latent fury bubbling forth, particularly from the second sentence, or chose to ignore it. Either way the document was "simply filed away."

The city's Manumission Society failed as well to protect Black graveyards from body snatching. The society had been formed in 1785, just three years before the Doctors' Riot, and boasted such luminaries as Jay, Hamilton, Duane, and Aaron Burr as members.

Jay, who had been outspoken in his disgust of slavery, owned five Blacks during his tenure as the society's first president. About a month before the council received the petition, the society had received complaints from Black leaders about bodies being stolen from the Negroes Burial Ground. After consideration, the society formed a committee "to inquire and collect proof of persons carrying away dead bodies from cemeteries." It seems no further action was taken on the matter.

However, even if the Common Council had ordered protection for the Black cemetery, and even if the Manumission Society had taken more drastic action, medical students would have had no shortage of other graveyards to plunder. By 1788, a total of sixteen churches of various denominations managed their own cemeteries. The city itself managed the potter's field across from the Negroes Burial Ground. Students from New York Hospital could have walked south a few minutes and been standing in the middle of Chambers Street, with a graveyard for Blacks on one side and for paupers and prisoners on the other.

About two weeks after the Common Council received the petition, resurrectionists (almost certainly medical students) stole a body from the graveyard of Trinity Church, at the corner of Broadway and Wall Street, a fifteen-minute walk from the hospital. The original Trinity Church, built in 1696, had been destroyed by fire in 1776. The second (and current) Trinity Church, built on the same site, wasn't completed until 1790, so the thieves might have seen the churchyard next to a partially completed building as a choice spot for their clandestine endeavor.

Church leaders responded quickly to the theft and offered a reward for the culprit: "Resolved that a reward of fifty dollars . . . be offered for discovery of the person, who, during the course of the last week, dug up a corpse, recently interred in the cemetery of Trinity Church." A church member, one Dr. Robert, matched that amount, bringing the total reward to $100, a sum equivalent to about six months' wages for a common city laborer.

The Trinity theft, as well as thefts from several other graveyards, prompted a signed letter to appear in the *Daily Advertiser* on February 16. It would be the first of several letters about body snatching printed in the paper. The *Advertiser*, helmed by Francis Childs, was the fourth-oldest daily newspaper in the nation and was widely read throughout the city. Childs anticipated that the letter might spark controversy, but he perhaps did not anticipate the extent. Thomas Gallagher, in his history of Columbia College, called *A Doctor's Story*, explained that Childs most likely found the letter shocking, "not in the sense that it told Mr. Childs something about the city he didn't know, but because it declared openly what was being discussed privately in tavern, home, and coffeehouse." He decided to publish the letter anyway but without a signature. "Mr. Printer," the letter began:

> *The repositories of the dead have been held in a manner sacred, in all ages, and almost in all countries. It is a shame that they should be so scandalously dealt with, as I have been informed they are in this city. It is said that few blacks are buried whose bodies are permitted to remain in the grave. And that even enclosed burial grounds, belonging to churches, have been robbed of their dead: That swine have been seen devouring the entrails and flesh of women taken out of a grave, which on account of an alarm, was left behind: That human flesh has been taken up along the docks, sewed up in bags; and that this horrid practice is pursued to make a merchandize of human bones, more than for the purpose of improvement in anatomy. In England, none but the bones of criminals are made use of for dissection. If a law was passed prohibiting the bodies of any other than criminals from being dissected, unless by particular desire of the dying or the relation of the dead for the benefit of mankind, a stop might be put to the horrid practice here; and the minds of a very great number of my fellow-liberated, or still enslaved Blacks, quieted. By publishing this you will greatly oblige both them and your very humble servant.*

Childs didn't have to wait long for the first response. Six days later, the newspaper published another letter, this one a sarcastic, caustic rebuttal of the first: "Great offense, it seems, has been given to some very tender and well meaning souls by gentlemen of the medical department for taking out of the common burial grounds of this city bodies that had been interred there. One in particular, whose philanthropy is truly laudable, has obtained a place for his moving lamentations in your useful paper."

The writer took direct aim at the first letter-writer and even bestowed on him a pseudonym: Humanio, Latin for humane. The writer signed his screed A Student of Physic and proceeded to reprimand Humanio for his views. "Kind and generous Humanio, thou standeth up in defense of the rights and privileges of the dead; and the dead, when they recover the use of their tongues, shall gratefully thank thee. At present they are dumb and silent."

Student of Physic went on to ask questions of Humanio and then to answer them himself. "And what say the living of thee, Humanio? I'll tell thee candidly, my honest fellow, that you are an unfit companion for them; that your head is too empty, and your heart too full; that through your excess of sympathy for the impassive bones of the dead, you have forgot the sore sufferings to which the living world is liable."

Student questioned, "Where is skill in surgery to be derived? From dissections only."

And "To whom would Humanio call for assistance should he snap his leg, or burst a blood vessel? Run, run to that barbarous man who has dissected most flesh and anatomized most bones."

Student closed his letter with "Adieu, Humanio! Betake thyself to the gloomy mansions of the dead, and bewail their wretched fate; it is an office that best befits your sympathetic heart!"

So began a point-counterpoint series of letters published in Childs' newspaper, arguing for or against body snatching, with Student of Physic at one point granting New Yorkers ever fresher fodder for their anger. One night he and several colleagues scurried under the cover of darkness to the home of someone who owned and managed a graveyard for Blacks on Gold Street. They banged on the man's door and ordered him to remain inside "at the peril of his life." The group then went into the graveyard and dug up a child's body. They were about to unearth another body, that of an elderly person, when the manager stormed out of the house and asked whether they "were not ashamed of their conduct!" Student of Physic denied any shame, saying he would think it "no crime to drag his [own] grandfather and grandmother out of their graves."

Shortly after that confrontation, a third letter appeared in the *Advertiser*, its circulation now expanding from the squabble. Published February 25, the letter bestowed a moniker on Student of Physic—"Mr. Abductio"—and facetiously labeled him "the hero at attacking dead bodies." The letter writer then threatened Abductio, warning "him, and his rash, imprudent and inconsiderate companions that they may not alone suffer an abduction of their wealth but perhaps their lives may be the forfeit of their temerity should they dare to persist in their robberies, especially at unlawful hours of night." The letter was signed, Humanio.

Letters from other interested parties began arriving at the *Advertiser*'s doorstep. One writer, "Philo-Smilio," responded on February 27 to one of Abductio's letters. Philo-Smilio, with tongue planted firmly in sardonic cheek, proposed that "funerals should be prohibited by law as useless and expensive and that the dead should be immediately surrendered at discretion to the surgeons." Philo suggested that a contest be held among medical students to see who could dissect the most bodies "in a given time" because "frequent handling of the knife and expeditiously cutting up a great number of bodies are the only means of acquiring a general knowledge of anatomy and skill in surgery." Besides prohibiting funerals and awarding prolific dissectors, Philo snidely proposed that "to check an inordinate appetite for carnage, which might probably arise among the young surgeons, I would punish, with the utmost severity, all those who . . . should willfully kill any person with a view to dissection, except those pronounced incurable by the faculty; in such cases it should be always permitted."

Another writer, "A Friend of Decency," clarified a key religious aspect of robbing the graves of Blacks, one that Abductio seems not to have considered. Friend of Decency admonished Abductio that "had he even exercised an ordinary attention to the operation of the human passions, he could not possibly have been ignorant *that the living may suffer in the dead*" (emphasis in original). The writer then added, "There is something peculiarly mean and dastardly in singling out the bodies of these people [Blacks] for the objects of such wanton outrages on the rights of sepulchre! After having stripped them, as far as possible, of all that ennobles human nature and gives to life its charms, is it not the extreme of brutality to take from them the gloomy consolation of looking forward to the grave, like the patient man in his distress, as a place where 'the wicked cease from troubling and where the weary be at rest.'" By the end of February the Black community began assigning armed guards to watch the burial grounds at night, perhaps inducing anatomy students to seek other graveyards for their prey.

The *Advertiser* published the last of Abductio's letters on March 1, or at least the last Childs chose to print. Perhaps no other letters arrived, or perhaps Childs, a member of the Manumission Society and avowed abolitionist-slaveowner, had tired of the arguments and passively put an end to the conflict. In any case, Abductio was granted the final voice in the matter.

Not only did Abductio fail to apologize for menacing the Black graveyard manager, he doubled down in his criticism of the poor man: "I shall not fill up your paper with a recapitulation of all [Humanio's] absurdities and blunders," he wrote, "but shall only . . . warn him not to be rash and imprudent, as again to attempt to espouse the cause of his fellow sufferers (for I take him to be some manumitted slave) without first applying for another quarters [sic] tuition at the free Negro school; that he may thereby be enabled to convey his meanings, at least in good, if not elegant, language."

Historians have tried to identify the letter writers, particularly Humanio and Student of Physic. Robert Swan, a specialist in the history of the Black community in New York City, wrote in a detailed article on the riot in *New York History Journal* that Humanio was "possibly a member of the Manumission Society or Trinity Church, perhaps both," and that "all of the explicit and implicit evidence and knowledge of surrounding circumstances and events suggests that Humanio was Scipio Gray." Gray was a free Black who, at some point after the war, had purchased a house and adjacent lot on Gold Street, about a half-mile walk from New York Hospital. "Gray obtained a lot of land," said Swan, "which, in agreement with other free blacks, was to be used as a private cemetery for certain members of the black community, probably Episcopalians. He engaged himself as its caretaker and guardian. By 1788, interred there were 'many bodies, adults and children.'" Swan further suggests that Gray probably wrote the petition to the Common Council and that he almost certainly wrote the letter calling Abductio a "hero at attacking dead bodies."

The identity of Abductio has never been confirmed, but he is generally believed to have been an anatomy student of Bayley named John Hicks Jr. His father, John Hicks Sr., was a physician who had worked with Richard Bayley and Samuel Clossy at a military hospital during the war. Bayley and Clossy had both recruited students to find bodies for dissection and had sometimes stolen bodies themselves. Student of Physic's full signature on his letters was "Student of Physic, Jun," with "Jun" being the common abbreviation at the time for Junior. In addition Hicks Jr. lived at 27 Broadway, just a few blocks west of Gold Street and Scipio Gray's cemetery. Swan said that Hicks "was indeed suspected of being both the writer of the letters and the man who threatened Mr. Scipio Gray, but it was never proved."

A nineteenth-century author named Joel Headley described the tenor of the time: "In the winter of 1787 and 1788, medical students of New York City dug up bodies more frequently than usual, or were more reckless in their mode of action, for the inhabitants became greatly excited over the stories that were told of their conduct. Some of these, if true, revealed a brutality and indecency, shocking as it was unnecessary. Usually, the students had contented themselves with ripping open the graves of strangers and negroes, about whom there was little feeling; but this winter they dug up respectable people, even young women, of whom they made an indecent exposure."

Most New Yorkers would have been at least somewhat aware of the issue. At the time, the city was home to between 25,000 and 30,000 people and as many as fifteen newspapers, including the *Daily Advertiser*, *New-York Packet*, and *New-York Journal*. Residents hungered for news that winter and spring, eager to hear more about state officials grappling with the ratification of the Constitution and looking forward to Hamilton's next Federalist Paper. Members of Trinity Church had been concerned enough to offer a bounty for anyone who could identify the men behind the theft of a body from their churchyard in February. No record exists of them ever finding the villains.

In mid-April, a group of students from New York Hospital ventured into a graveyard and snatched the body of a White woman, brought it to Richard Bayley's anatomy lab in the otherwise empty hospital, and there began to dissect it. That act, commonplace though it was to the students, was the beginning of what has been called the "largest riot of the late eighteenth century to regulate communal morality."

CHAPTER 10
THE OBNOXIOUS DR. HICKS

Few contemporaneous sources about the doctors' riot have survived, and those that have lack substantive detail. For instance, the name of the woman whose body was stolen has not survived, nor have her age or details of her or her family's life. Precious little is known about John Hicks Jr., whose actions started it all, or of the physicians and students present in the lab at the time. Stories of the riot itself have been told and retold with considerable inconsistency, like eyewitness accounts after an auto accident. However, a thorough review of contemporaneous articles, facts gleaned from well-cited accounts by historians, and a solid dose of commonsense reasoning can provide a balanced and detailed, if at times slightly fictionalized, account of the riot that sparked decades of similar unrest throughout the country. Herewith is that account.

THE PEWS AT ST. PAUL'S WERE FILLED WITH CHURCHGOERS THAT day, Sunday, April 13, 1788, as they usually were. Governor George Clinton was there, as were Mayor James Duane, Mr. and Mrs. John Jay, Dr. and Mrs. Samuel Bard, Alexander Hamilton, and Mr. and Mrs. Brockholst Livingston, the same Brockholst Livingston who helped Hamilton defend Joshua Waddington in his case against Elizabeth Rutgers. After the service, everyone said their goodbyes and departed on foot or riding in a carriage.

A few hours later, seven blocks north of the church, a small group of boys headed to a grassy area behind the nearly empty, still under construction New York Hospital. The hospital was a three-story, H-shaped, gray stone building with a basement. The building had tall, multipaned windows all along the first and second floors and somewhat shorter windows along the third floor. Windows along the exterior at ground level provided sunlight to basement rooms. Bayley's anatomy lab was located on the first floor in the back of the building and looked out over a grassy area, secluded enough that it had become a favorite venue for duels.

The boys didn't bother with any of that, they just enjoyed playing somewhere they wouldn't be bothered by adults. Anyone who happened upon them that afternoon would have seen a bunch of boys being, well, boys. At some point something in a window above them caught their attention. They stopped playing and moved toward the building. It was a severed arm hanging from the window.

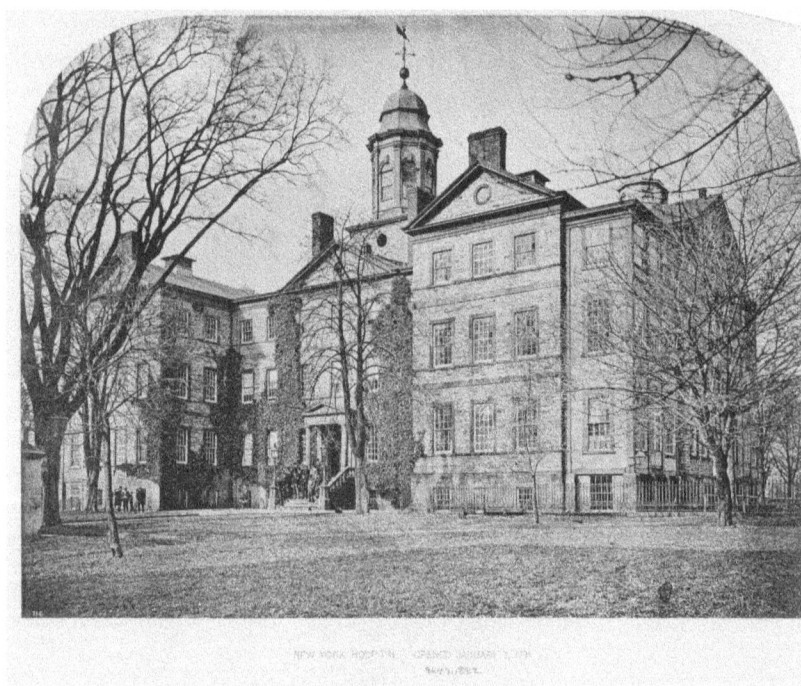

Photogravure of the original New York Hospital, 1791. COURTESY OF THE MIRIAM AND IRA D. WALLACH DIVISION OF ART, PRINTS AND PHOTOGRAPHS: PICTURE COLLECTION, THE NEW YORK PUBLIC LIBRARY.

The boys were dumbstruck. They could have run away. They could have just gone on playing, but they were boys, and boys did what boys do. They investigated.

Workmen had left a wooden ladder leaning against the building, so a few of the boys raised the ladder to the window. One daring lad jumped on the ladder and began to climb, egged on by those on the ground. Nearing the top the boy tried not to brush against the gray and mottled arm, but he did take a long look at it. Then he peered into the room. A skeleton stood against the far wall. He made out several large tables, and on a table in the middle was a cluster of older boys—men, really—looking at him, *directly at him*! As he took in more of the room, he noticed a man standing next to the table talking to the boys. On the table was, wait, what is that? *O! A woman, naked! And the man is cutting into her!*

One of the older boys, John Hicks Jr., decided to have some fun with the interloper. "That's your mother's arm," said Hicks, "I just cut it off!"

The boy's mouth dropped open. His mother actually had just died, and after the man's words had penetrated the boy's shock, he scurried down the ladder to tell his friends. The other boys were shocked as well. Then they all sprinted home to tell their parents.

The boy who had climbed the ladder burst through the front door of his home, ran to his father, and blurted out the nearly unimaginable story. The boy's father was a Freemason who had been working on a building on Broadway. He most likely belonged to St. John's Lodge on Ann Street, which originates at Gold Street, not far from Scipio Gray's house, and terminates at Broadway, across from St. Paul's Church. His father initially wasn't sure about the story, but at some point, he realized that his son must have seen someone in that room and that he had better check his wife's grave to be certain. He had heard stories about unscrupulous men digging up graves and about doctors cutting bodies open. The more he thought about doctors doing that to his beloved wife, the angrier he became.

The man grabbed a shovel and the mallet he used at work and headed out the door. He sped to the homes of a few fellow workers and neighbors, telling each the story his son had told him and saying that he was going to destroy the lab. After hearing the story, the men proved just as eager as the father to rain hell on those damned doctors. Racing to the gravesite of the boy's mother, the men found the grave uncovered and the body gone. Now incensed, the group stormed off toward the hospital.

By the time the crowd neared the back of the hospital, it had reached a total of perhaps two to three dozen men. The men hustled to the building, and as they drew closer, they could make out the arm still hanging there, a gruesome spectacle. The group followed the grieving father as he ran up the stairs to the rear entrance and entered the building. When they identified the anatomy classroom, the boy's father yanked open the door, and he and the others rushed in, their faces flushed with anger.

After the boy on the ladder had climbed back down, Hicks and a few other students ambled to the window and watched the boys as they scattered. The professor, Richard Bayley, quickly called the students back to the dissection table to get back to work. Bayley and his students were dissecting the body of a White woman, "a very handsome and much esteemed young lady of good connections." The body had already lost one arm and was in the process of losing another. It was not the only body being dissected, either. A short distance away, Wright Post was hovering over the body of "a young gentleman from the West Indies." His and the woman's body had both found their way to Bayley's classroom over the last couple of days.

Bayley, Post, and their students soon lost themselves again in their dissection duties, perhaps carefully separating the median nerve from the biceps muscle in the upper arm, piercing the small intestine to reveal the circular folds inside, or incising the largest artery in the body, the aorta, and slicing down through the thick wall of the left ventricle to view inside the heart. None of the medical men thought again about the lad at the window. The students were focused. They were learning to become physicians, to heal the sick and injured, to adhere to the rule *Primum non nocere*—first, do no harm. They were too busy to hear robins in the trees, the low rumbling of carriages passing along Broadway, or the thrum of angry voices in the distance.

One of the students gradually became aware of people approaching. He stepped to the window and saw a swarm of men moving quickly across the lawn. He heard the men shouting and called out to Bayley, who went to a window and saw for himself what was unfolding. Bayley immediately realized what the men wanted. He turned to his students and told them to run, to get out as fast as they could, that they were in danger from the men outside. The students scattered. One hid in a chimney, and others hurried to escape the building. Post and a few students stayed behind to guard the specimen collection, which had taken Bayley years to build.

The men rushing into the classroom must have been appalled by what they saw. Two naked bodies, one female and one male, both lying face up on tables. The men soon discovered a third body "boiling in a kettle," the typical technique for melting away muscles and other tissues so that only a skeleton is left. Bookcases along the walls displayed jar upon jar of human organs and tissues. A brain here, a stomach there, an entire loop of small intestines in one jar, a pair of eyes in another. A dried skeleton was hanging on a hook on the wall. A liver, heart, pancreas, and large intestine, all recently removed from the West Indies gentleman, sat atop a table in the corner, glistening and gray in the afternoon sun.

The boy's father was drawn to the body of his wife, laying on her back in the middle of the room on a cold table, her arms torn off, her abdomen slashed open, her eyes vacant and fixed. He could barely look at her, yet he couldn't look away. Finally, he found a cloth off to the side and covered her body. The other men started smashing specimen jars on the floor, glass shattering and preservative fluid spraying everywhere.

They found surgical instruments splayed around where the students were sitting—hand drills for opening holes into the skull to relieve pressure, bone saws, scalpels of all kinds, various sizes of sickle-shaped amputation knives, lancets, razors, pliers, and a few small hammers. These too were slammed on the floor or pocketed. There was a large bucket on the floor near each table for collecting fluid, and sheets of paper covered with illustrations of body parts were tacked to the walls. The men tipped over the buckets and tore the papers off the walls and ripped them up. They destroyed the skeletons they found and damaged or destroyed books and student notes.

On shelves and tables were bellows and beakers and flasks in a splendid assortment of colors and styles, all for mixing dyes and chemicals. Many of those, too, were broken in the melee. Some of the men found another room that contained specimens and equipment and destroyed them as well. Whether the men knew how much the equipment cost, where the specimens had come from, or how long it had taken Bayley to collect them is unknown, but it didn't matter. They were out to hurt whoever had taken the man's wife from his grave and cut her up. They were furious at the doctors—all doctors—and sick of bodies being stolen from what should have been their eternal resting place.

A letter to the editor by "Order, Decency, Justice, and Peace," writing in the *American Herald*, a Boston publication, provided an interesting take on the ruckus. "It is true," the writer said, "the faculty must sometimes dissect in order to teach their pupils and make such preparations as may, in various cases, prove beneficial to mankind. But these great ends have always heretofore been answered by such prudent and cautious methods as have never alarmed the minds of the people." The writer went on to explain that "one or at least two subjects are abundantly sufficient" for anatomy study and that criminals and "some forlorn corpse to which none claimed connection" were quite acceptable for the purpose. Yet here were honorable people having to deal with the bodies of their loved ones being stolen. The writer continued, "There never was, perhaps, in a well-regulated society a great outrage deliberately and frequently committed, nor the feelings of a whole city more impudently insulted, by unprincipled youths or some of the imprudent masters."

The honorable people plundering the hospital discovered Post and a few students, including young David Hosack, protecting the specimen collection. The rowdies beat them all, pounding them with fists and feet and whatever objects they carried. As luck would have it, the mayor, James Duane, suddenly arrived, bringing with him the sheriff and a few rattle watchers, regular civilians serving as part-time, amateur police officers. The city lacked any kind of organized law enforcement until the 1820s, using instead a system of watchmen carrying green lanterns and wooden rattles. Teams of rattle watchers would patrol streets to discourage thieves and warn residents of fires. It might have been one of the watchers who noticed the mob and alerted the mayor. The sheriff carried with him a writ of *mittimus*, a court order allowing him to place his prisoners, the students, in jail for their protection. His arrival no doubt saved the lives of Post and his students.

Duane ordered the rioters out of the hospital, an order they reluctantly obeyed. The boy's father and a few other men carried the remains of the wife, along with the man from the West Indies, down the stairs and outside. They placed the bodies in carts to be carried to the cemetery, there to be "interred the same evening in triumph." Denied an outlet for their wrath, the crowd dispersed for the night. Vengeance would have to wait.

Chapter 11

HUNTING FOR THE CURSED DISSECTORS

STORIES ABOUT THE PREVIOUS DAY'S EVENTS SPREAD THROUGHOUT the city overnight, inflaming anger in an already affronted citizenry. Word also spread among city officials and dignitaries who might be able to help calm the crowds. Duane contacted a few of his friends to help contain the city's outrage, men who fortunately lived only a short distance from his office. Duane himself lived well outside downtown, two miles northeast of the hospital at Gramercy Farm, but he worked most days at the mayor's office in Federal Hall, located at 26 Wall Street—a quick two-minute walk to Hamilton's house, at 57 Wall Street. Jay, serving at the time as Congress's secretary of foreign affairs, lived a bit farther away from Hamilton's house, about a four-minute walk, in a "large, square, three-story house of hewn stone, as substantially built within as without," and "durable, spacious, and commodious." Clinton lived at the Governor's Mansion, near the corner of Pearl Street and Maiden Lane, also a four-minute walk to Hamilton's.

Duane and the others might have hoped that the previous day's anger had resolved on its own, but if so, they were badly mistaken. In fact, the situation was made worse. Much, much worse. In an astounding show of either gall or dreadful timing, that evening—the same evening the men were burying two bodies rescued from the scalpel—yet another body was stolen from Trinity's churchyard.

Two men, a medical student named Isaac Gano and a musician named George Sweeney, dug up the "dead body of a certain white woman." Both men were shortly thereafter indicted for illegally entering a church cemetery. The men "then and there unlawfully did dig up, take out of the same coffin, and carry away" the woman's body. "If the indictment was based on accurate evidence," said Ralph Victor, an anatomy professor at the University of Rochester, "one must admire their audacity. Since Gano was a medical student, his action might be understandable on the ground that the school had just been depleted of all anatomical material."

Despite the indictment, no public announcements were made. Said Victor, "No New York newspaper published anything about it, although the papers of that time are full of indictments against other persons for different reasons. Also this case is not indexed in any digest of American court procedures." It seems likely that city officials squashed the story for fear of instigating another demonstration at the hospital. Just as likely, however, is the possibility that Gano was not affiliated with Bayley's class at all and so wouldn't have known about the goings-on in the anatomy classroom that afternoon. Regardless, Gano and Sweeney faded into the background.

Bayley and Post, however, did not. Post had already been sequestered at the prison with Hosack and the others, a group now guarded by eighteen armed rattle watchers. Bayley, who had escaped the mob at the hospital, made his way during the night to the jail and joined his protégé and students.

Early the next morning, April 13, a crowd began to collect at the hospital. Several hundred citizens joined in, and the throng began to head south on Broadway. They had heard that a well-respected physician, a Dr. John Cochran, might be protecting Hicks. Cochran was the fourth in a series of medical directors for the Continental Army, serving right after William Shippen. Now retired from practice, Cochran was living across Broadway from Trinity Church. The rioters pushed their way through a funnel-shaped section of roadway between Broadway and Chatham Row. They stormed past the John Street Theater, the birthplace of American Theater. They passed the canopied Oswego Market on Maiden Lane, a busy marketplace during the day. As the mob marched toward Cochran's house, they were met by what must have been an intimidating contingent of dignitaries: Mayor Duane, Governor Clinton, and Robert R. Livingston, the state's attorney general, along with a number of other prominent New Yorkers. Duane addressed the crowd and "endeavored to dissuade them from committing unnecessary depredations." They promised them "every satisfaction which the laws of the country can give."

Some people seemed satisfied and left "peacefully with the Mayor's assurance," but most remained. The protesters sought revenge on Hicks and the other students and physicians, and they wouldn't stop until they succeeded. Duane and his distinguished partners could do little to stop them, though they continued to try, walking along and speaking with this group or that as the entire mob moved steadily toward the good doctor's house.

When the crowd finally reached Cochran's home at 97 Broadway, the doctor let a small group of rioters into his home—he could hardly do otherwise—and watched as they searched his abode from "cellar to garret with no success." So intent were they on finding Hicks that they "even opened the scuttle and looked out upon the roof," failing to perceive the terrified Hicks hiding behind the chimney of an adjoining home. On leaving the house empty-handed, the men told the crowd that Hicks was nowhere to be found, at which point a number of rioters took their leave and returned home. The remainder continued the search.

Similar scenes played out at "every physician's house in town." The mob's quest brought them to the home of Charles McKnight, who gave lectures at Columbia College; Nicholas Romayne, who used rooms at the Almshouse to teach anatomy; and even the esteemed Samuel Bard. No damage was done to any of the homes. "It was a singularly well-behaved mob," wrote author Joel Headley, somewhat inaccurately, "and when they had made the round of the houses, gradually broke up into knots and dispersed." The mob might have broken up a bit, but it nevertheless marched north on Broadway and crossed over at the intersection with Barclay, heading for Columbia College. The crowd stormed across the lawn in front of the college and gathered at the front entrance.

The scene proved reminiscent of an earlier time, when Alexander Hamilton stood on the steps of then-King's College and implored hundreds of the protesters gathered there to remain calm. Sources differ on whether Hamilton performed a similar service at Columbia College, but it seems likely that he did. He was a former student at King's and had taken an anatomy course with famed instructor Samuel Clossy, so he would have wanted to protect the school. He certainly would have heard about a mob attacking New York Hospital, and such was his sometimes vainglorious attitude that he probably did trek to the college to help. If he did indeed stand on the front steps of the college and try to steady the crowd, it didn't work. The mob pushed past him, too angry for oratory. The protesters barged through the front doors, shocking students in hallways and classrooms into stunned silence. They "swarmed without opposition throughout every part of" the building, finally departing in frustration. By that time a number of citizens who had heard about the stolen bodies and the cutting up of that "very handsome and much esteemed young lady" found the mob on Broadway and joined in.

As the crowd grew, a silversmith named William W. Gilbert was called to the prison. Gilbert was an alderman at the time and had been summoned to take a deposition from Richard Bayley, who wanted an official document denying his involvement in any crime that might have occurred at the hospital. Published the following day the deposition read:

RICHARD BAYLEY, of the City of New-York, being duly sworn, deposeth, and saith that he hath not directly or indirectly had any agency or concern whatsoever in removing the bodies of any person or persons interred in any churchyard or cemetery belonging to any place of public worship in the said city; and that he hath not offered any sum of money to procure any human body so interred for the purposes of dissection. And this deponent further saith that no person or persons under his tuition have had any agency or concern in digging up or removing any dead body interred in any of the said churchyards or cemeteries to his knowledge or belief.

Bayley cunningly left unsaid whether he might have obtained bodies taken from cemeteries that didn't belong to a place of public worship, such as the Almshouse graveyard or the Negroes Burial Ground next door. He also added the convenient codicil, "to his knowledge or belief." Why, no, he had no idea what his students were doing or where those cadavers had come from. In truth, he certainly knew what his students were doing; he had performed that task himself, after all. Even if he didn't know they had stolen a White woman's body, he probably would have welcomed a female in her thirties as a subject.

John Hicks would write a similar abnegation the day Bayley's denial appeared. Hicks sat for a deposition taken by Alderman John Wylley. How he arranged the deposition and where it took place has been lost to history, but surely Hicks felt comfortable enough to set up a meeting with Wylley and to believe he wouldn't be tricked into being captured. In the deposition, Hicks claimed that, like Bayley, he, too, had no "agency or concern" about any theft of bodies. Even more outlandish, however, he claimed that he "hath not been within the walls of the hospital in this city since the year 1783." That is, he added, not "until Sunday last by mere accident, on account of a large number of people having collected at said hospital."

If anyone in the mob on Broadway had heard about either Hicks' or Bayley's denials, it probably wouldn't have made any difference. Leaders at the front of the mob cried out, "To the jail! To the Jail!" The mob moved nearly as one northward shouting along the way, "To the jail!"

The jail, called Bridewell Prison, was used during the Revolutionary War to house American prisoners-of-war and thereafter as New York City's main jail. An imposing stone structure built in the Georgian style of architecture, Bridewell contained a dungeon-like basement with nine-feet-tall brick archways and walls two feet thick.

BRIDEWELL, PARK, N.Y. 1789.

Lithograph of Bridewell Prison, by George Hayward, circa 1855.

Thick oaken doors separated holding areas within the basement. The main door to the building had been breached in 1764 by British troops, angry that Major Robert Rogers, a popular but profligate officer, had been jailed by his creditors. The troops used axes to break down the door and then released Rogers, who, a year later, would be evading creditors once again, in London. After that breach, the prison's door was replaced with a stronger door and lock. The cost of that replacement would pay off a quarter-century later.

The mob heading north on Broadway increased throughout the day to "an alarming size," according to one newspaper report, prompting Mayor Duane to order militia to repair to the jail immediately to protect the prisoners. He asked for volunteers for the mission, said historian Robert Swan, but with many men "sympathetic to the [mob's] cause, few mustered for duty." Finally, a dozen or so armed militiamen gathered on Broadway, south of the mob, at around 3 p.m. and marched toward the jail. They caught up with the mob and passed through mostly unmolested, but then a few rioters began pummeling the conscripts with stones and handfuls of dirt. The men took off running for the jail, setting up a defense when they arrived. A few men stood guard at the rear door, bolted from the inside, and the rest took their place at the front, where they expected the mob to storm.

Not long after the mob assaulted the first group of militia, another dozen or so militiamen bound for the jail came upon the mob. This time, the mob was in no mood to be interrupted. Rioters quickly swarmed the troops and beat them. Without orders to use force unless the rioters entered the prison, the soldiers refused to fight back. The mob seized the men and destroyed their rifles.

Thrilled at their victory over the troops, the mob moved even more rapidly toward the jail. When they finally arrived, they tore down the picket fence surrounding the building. They screamed over and again, "Bring out your doctors!" The militiamen yelled back that the doctors would not be coming out, nor would the rioters be let in. A few of the strongest militiamen inside guarded the door, sure that the mob would soon try to force it open. The door was constructed in the common stile-and-rail design, with vertical stiles and horizontal rails framing a half-dozen panels, considerably stronger than the jail's first door. The rioters quickly tired of threatening the guards with their wrath and instead did exactly as the guards had predicted, thrusting themselves against the door to bust it open. Having torn down the gallows and whipping post on the grounds and broken them apart, the rioters now used the largest beams as battering rams against the door.

Inside, the defenders pushed back, heaving themselves against the inside of the door. The lock and hinges held tight, as did the men. The lock was a rim lock, a precursor to the modern deadbolt lock. Rim locks were commonly used for exterior doors on more expensive homes and buildings, the cost precluding their use in typical houses. Each rim lock made during that time was handcrafted and had its own particular keyhole shape, which allowed one particular key to enter. Once inside the lock, the key would turn the lock box, attached to the inside of the door. With doors several inches thick, keys were typically six or more inches long. Without that key, usually held by the main jailer, the lock could not be opened.

Even against a battering ram, the lock and hinges held steadfast. Unable to break the front door down, rioters started pelting the building with rocks and brickbats. Windows shattered. Shards of glass splintered all around the guards and students inside. Ransackers used planks of picket fence to pry open the windows, but the militiamen inside slammed their rifles on any hand or arm that came through. "Every picket fence within blocks of the jail," said historian Jules Ladenheim, "was destroyed to provide weapons for the mob. Great as the commotion was, however, up until this time no lives had been lost on either side except one." A rioter who gained entry into the prison through a broken window was bayoneted by a soldier inside and killed. Some of the rioters looped around the jail and attacked the doors there. Those doors were bolted shut from the inside and could not be unlocked from the outside, nor could they be broken down.

Guards inside the jail bellowed to one another and to the rioters outside. The crowd outside screamed obscenities and commands back at the guards. Now and then, the mob quieted, with the leaders who had emerged during the day talking to one another, determining strategies going forward. During those breaks, the Bridewell guards would prepare for another assault, with each onslaught repelled by the defenders.

The mob, as large and angry as it was, did not burn down the jail. Protesters did not insert gunpowder into the keyhole and blow up the front door lock, nor did they shoot into the building. They did not burn or damage Dr. Cochran's house. They, in fact, treated the doctor's house with respect, searching it for student Hicks but not flipping over chairs, breaking windows, or smashing Cochran's furnishings. Some protesters "did much mischief and damage" to the homes of other physicians, and apparently several people "got much bruised and wounded in the fray," but otherwise, the rioters proved rather docile. The mob also hadn't set fire to the classrooms at New York Hospital. Only items in the anatomy rooms were damaged. No other classrooms were touched, nor were any administrative areas. "Throughout, the rioters maintained a sense of purpose and limited their violence against both persons and property," explained historian Paul Gilje. "Because the building itself was not associated directly with the affront to public morality, it was not damaged seriously. Moreover, although the rioters captured and abused several medical students, they surrendered their prisoners to government officials on the promise that there would be legal action instituted against the doctors."

A third group of militia arrived at the prison around dusk. As they made their way along, the crowd hissed and jeered, niggling the men with taunts and threats. Tensions surged, anger turned to fury, and the mob's behavior bordered on hysteria. Continued violence seemed assured.

CHAPTER 12

BATTLE OF BRIDEWELL

Word spread rapidly about the riot at Bridewell and the medical men trapped inside. Mayor James Duane and Governor George Clinton had been scurrying about all afternoon in search of volunteers to aid the militia and alerting as many city officials as they could find. When Matthew Clarkson heard about the mob, he rushed to the home of John Jay. Clarkson was Brigadier General of the militia of two counties, Kings and Queens, and commanded two thousand soldiers. He had sent a few handfuls of troops already to guard the students and, hearing that the size of the mob had far exceeded his estimates and that soldiers at the prison were exhausted and grossly outmatched, he ordered additional troops to the prison. Clarkson then hustled to Jay's house to warn him.

As Clarkson reached the Jay home, he ran up and burst through the front door. "My God, Jay!" Clarkson is said to have cried out. "The mob's surrounding the jail! They mean to break in and tear the doctors to pieces. Sword! I need a sword!" Jay bolted up the stairs and, moments later, darted back down. "Handing Clarkson one sword," recalled Jay's wife, Sarah, "to my great concern, he [Jay] armed himself with another and went towards the jail." Jay and Clarkson jumped into Jay's chariot, a smaller and lighter type of horse-drawn carriage, and drove off for the prison. By the time they arrived, Clinton, Duane, the famous Baron von Steuben, and even the great Alexander Hamilton were already standing near the front door, pleading with protesters to stand down. Also present was General John Lamb, who had commanded artillery troops at Yorktown, and more than a dozen other city dignitaries.

Portrait of Matthew Clarkson by famed artist Gilbert Stuart, circa 1794. Clarkson's daughter, Mary Rutherfurd, would marry Jay's oldest son, Peter Augustus, in 1807. COURTESY OF THE METROPOLITAN MUSEUM OF ART. BEQUEST OF HELEN SHELTON CLARKSON, 1937.

Clarkson and his sword strode ahead of Jay as they moved to the front of the crowd. The brigadier general, age thirty, was a handsome man with dark brown eyes and a stern look. He could grow a beard quickly but chose not to wear one. He was not a tall man, nor was he broad-shouldered, but his stern countenance meant business. He barked and shoved his way through the rabble, making a safe path for Jay, at forty-eight the elder statesman. The protesters allowed them to pass unmolested. At the door now were a handful of imposing, illustrious, and highly influential men standing against as many as five thousand enraged citizens who had yet to infiltrate the jail.

The troops Clarkson had ordered to the jail finally arrived. "We marched up to the jail," said one soldier, writing to a friend in Boston in one of the only eyewitness accounts ever found, "and the mob waited for us until we were within ten paces of the door. Our orders were not to fire. The mob were of the opinion that we dare not fire, or if we did, it would be over their heads."

The stage was set. Here was an aggrieved throng spread out in front of the prison, all clamoring for revenge. Many of the protesters, like the father who had seen his wife's body splayed open on a cold, unforgiving table, had felt the pain of having the body of a friend or loved one stolen from their grave. They knew the sadness of death and the outrage of discovering how their loved one's earthly remains had been defaced by people they didn't know or care about. How could they feel peace knowing how those remains were used? Why couldn't the students learn on something else, someone else, someone who didn't matter so much?

And here were about two dozen militiamen armed with rifles, swords, and clubs flanking perhaps twenty pillars of society and protecting the prisoners sequestered inside. A few of those prisoners had dedicated their lives to helping the sick and injured, had traveled overseas to study with some of the best physicians and surgeons in the world and returned to their homeland to set up their own medical practices. They had treated gunshot wounds and broken bones, applied poultices and lanced boils, delivered babies and helped ease dying patients on their final journey. Most of the students inside were just teenagers starting their odyssey into medical practice. They were mere striplings exploring the intense complexities and never-ending miracles of the human body. They were just boys, really, who had been doing exactly what their teachers had asked them to do: find cadavers to study the organs and sinew inside every human. How else were they to learn? How else could their professors teach them? How else could they become the doctors they wanted to be and to save the lives of others, as their professors had done before them?

Two opposing factions were now lined up within inches of one another, one determined to punish the people inside and the other equally determined to prevent it. The fury driving the mob stemmed almost exclusively from the theft of White bodies from churchyards reserved for White parishioners. No fusses from the non-Black citizenry had arisen before this, when Black bodies had been taken from the Negroes Burial Grounds. No fusses had come from the theft of bodies of the poor, indigent, and imprisoned in the city's potter's field. It was only after White New Yorkers realized that bodies of their own kind had been taken that protests had started.

Whether there were Black rioters among the throng has never been determined for certain. New York historian Robert Swan has outlined his reasons for believing that at least some Black citizens took part in the mob, writing:

> [Chief] Justice [Richard] Morris implied that the mob comprised all ranks of society, from the "vulgar" to those of refined sensibilities. Certainly, blacks—many who were light-complexioned and could pass for white—were interspersed within this throng. Three were arrested but cannot be identified. Between three and six deaths were reported; two were described as only a cartman and a young man. One black death occurred, that of a boy slave attending his master, William S. Livingston, who was defending the gaol house. Importantly, the riot began on Sunday afternoon. A mob of this size on the Sabbath probably did not emanate from the workplace, but more likely from church congregations,

perhaps instigated by and including ministers still intoxicated with Revolutionary fervor.

Contemporaneous newspaper articles, however, fail to mention the presence of non-White protesters. They mention Black citizens only in relation to the theft of bodies from the Negroes Burial Ground, "which has been the cause of loud complaints . . . for some months past." Those complaints, to the city's Common Council, had been left unanswered. When newspapers discussed the theft of Black bodies, they did so consistently outside the realm of so-called decent society. "The interments not only of strangers," claimed one article in the *New-York Packet*, "and the blacks had been disturbed, but the corpse of some respectable persons were removed." No matter the mob's makeup, the intensity of the anger rioters felt, and the complete frustration of negotiators unable to calm the crowd finally melded and passed an emotional and physical Rubicon.

Dusk had come and gone, and rain had started to fall. Lighted torches appeared throughout the crowd, tossing a golden glow over faces contorted in rage. Mere feet separated the madding horde from the unarmed authorities. The militia stood alongside the dignitaries, ready to fight back when ordered. Just then, about one hundred troops under the command of General John Armstrong stormed into the jail yard. About half of the troops carried rifles or pistols, and the rest carried clubs or other weapons. Along with Armstrong's cohort, another fifty members of a light horse regiment under the command of someone named Stokes also arrived.

The reinforcements spurred the crowd to a frenzy. They started throwing stones, rocks, brickbats, shanks of wood, anything they could find and hurled them at the troops and city officials near the jail. In seconds, a brick smashed Jay on the forehead, slicing it open. He fell, bleeding, to the ground. The dignitaries instinctively raised their arms to protect themselves, but it was of little use. They felt the sting of dirt, rocks, and brickbats slamming into them, tearing their clothes and cutting their flesh open. Duane was struck in the head by a projectile.

Steuben screamed at the rioters to stop, but the din of wrathful voices overpowered his words. He had been in combat many times before. A soldier since the age of seventeen, he had been wounded more than once during the Seven Years' War. He would not back down, nor would he tolerate for long remaining defenseless. Steuben had never been a fan of mobs, writing to a friend about the rioters of Shays' Rebellion two years before, "This mob consisted of men of the vilest principles, desperate in their fortune etc.—therefore a disgrace to human nature." However, Steuben also intimately understood the heartbreak of war, the trauma of shooting someone armed only with a rock or a stick, and the horrors experienced by anyone in the midst of war. So he had been pleading with the governor to hold his command to fire. However, when a brickbat struck Steuben himself in the head, cutting him just above the left eye, whatever benevolence he had been feeling dissipated. The war hero fell to the ground and yelled, "Fire, Governor! *Fire!*"

Clinton repeated the command to the troops, screaming "*Fire! Fire!*" Fire they did, first over the protestors' heads. After that failed, the troops leveled their muskets at the crowd and fired. Three protesters, possibly four, fell dead straight away. Others stumbled backward, wounded but alive. Clinton ordered some of the militia to charge the mob in the middle. "We drove them with our bayonets," said one soldier present at the prison, "as far as the Brick Meeting House. They collected on our flanks and in our rear, upon which we fired."

The crowd then swarmed around either side of the soldiers "in such a manner that we were forced to retreat." The troops that had arrived behind the mob were at that point positioned nearly opposite the troops moving back toward the prison, and they too were firing into the crowd. "It was so dark and rainy," according to the soldier's letter, "that we could not distinguish an object twenty yards."

At some point the mob turned away from the prison and began forcing their way down Broadway, their attempt to break into the prison having failed. The crowd began dispersing as it moved. Some protesters remained at the prison for a time, alone, the militiamen having been ordered to collect at the side of the prison after the shooting had ended. Eventually the rioters, tired and wet, left the area for home or their favorite pub. The madness was over.

Hundreds of stones, planks, and bricks had been thrown, and perhaps fifty or more rounds had been fired. Jay and Steuben lay wounded on the ground. Armstrong had been badly bruised in the leg, but he could walk. Numerous other dignitaries had been wounded as well, as had a number of militiamen. One rioter later died from his wounds.

Jay was taken to the almshouse first, and then to his home. One of Jay's biographers, Walter Stahr, wrote that "Sarah was appalled to see her husband returning 'with two large holes in his forehead' and feared that some permanent damage might have been done. Sarah wrote to her mother that the family doctor "immediately ascertained his wounds, and to my unspeakable relief pronounced that there was no fracture." The doctor dressed the wound and then, according to one source, bled the injured man, as was the custom. Jay survived both insults, lying in bed for several days with blackened and swollen eyes, "vast pain from his neck and shoulders," and considerably less blood than he had started the day with. The militia stood guard around Bridewell and patrolled the city the rest of the night. They encountered no further violence.

The next morning, Tuesday, April 15, a brigade of troops under the command of General William Malcolm and an artillery regiment under the command of Colonel Sebastian Bauman arrived at the prison. Malcolm was a Scot who had immigrated to America in 1763 and in early 1776 was appointed the first major of the Second Battalion of the New York Independent Companies. Bauman had started the war as a captain in a New York artillery unit and risen to the rank of lieutenant colonel in the Continental Army. Both men had served under General Lamb, who no doubt had contacted them sometime on Monday and requested their troops to aid in the prison's defense. The new troops guarded the prison and marched or rode throughout the city as a show of force, determined to prevent more bloodshed.

Throughout that evening's skirmish at Bridewell and well into the night, small bands of rioters here and there continued to search for doctors and medical students. To avoid capture, physicians would "slip out of windows, creep behind bean barrels, crawl up chimneys, and hide beneath feather beds," said author Jules Ladenheim. "The grave gentlemen of the healing art were forced to flee in dark places, like haunted rebels or persecuted prophets, for three days and three nights."

Protesters mistakenly attacked Sir John Temple's house believing that Temple was a physician. He was, in fact, a diplomat serving as the first consul general for Great Britain, a position in which he aided the passage of people, goods, and services between America and Britain. The mistake apparently arose because "Sir John" sounded strikingly similar to "surgeon," which proved enough of a reason for rioters to descend on the diplomat's house. It was only with the "greatest difficulty" that Temple was able to persuade the protesters that he was not a surgeon and that they should not destroy his home.

Clinton ordered that a squad of soldiers remain at the prison throughout the night and for a few days thereafter. No mob collected again, prompting the *New-York Packet* to express that "the peace of the city is once more restored" and that "respect to the magistrates and obedience to the laws are the principle securities for the safe and quiet enjoyment of life, liberty, and property. But from mobs, riots, and confusions, 'may the good Lord deliver us.'"

The riot at Bridewell Prison had been smothered, but anger seething in the public's consciousness had not. Another riot against anatomy students occurred just months after the New York riot. A German surgeon and inventor named Charles F. Wiesenthal had begun teaching anatomy around 1770, in a two-story building he had built behind his house on Gay Street in Baltimore, Maryland. The body of an executed murderer, Patrick Cassiday, had been given to Wiesenthal specifically to be dissected by his students. Some Baltimoreans took exception to the dissection. A mob formed two days after Christmas, stormed the building, and demanded Cassiday's body. The anatomy students, greatly outnumbered, allowed the mob to remove the body without objection. No one was injured.

Baltimore became the site of another anatomy riot, this time in late 1807, when a mob attacked a building called Anatomy Hall. The building, located at the corner of Liberty and Saratoga Streets, had been specially built by a physician named John B. Davidge to teach anatomy through lectures and dissections. A week after Davidge's first class, a mob attacked the building and burned it down.

The medical college at Yale University likewise experienced a resurrection riot. The college at the time was housed in a remodeled hotel at the corner of College and Grove Streets in New Haven, Connecticut. Students of Jonathan Knight, the school's first professor of anatomy and physiology, had snatched the body of a female on the night of January 7, 1824, and brought it back to the school. The body was that of nineteen-year-old Bathsheba Smith, daughter of a West Haven farmer. "We know little about this ill-fated lady," wrote author Hannibal Hamlin, "except her fascinating name. A newspaper version of the outrage describes Miss Smith as 'a respectable young female.'"

Constable Erastus Osborn and a group of men had heard about the snatching. They investigated the gravesite and discovered it empty. "It was opened [with] a flat stone found on the head part of the lid," said Osborn. "The stone was removed, and the lid found broken in and the body missing." Osborn led the group to the most logical place to look for Smith's body—the medical college. The men "searched top to bottom" and were about to give up when Osborn decided to check the basement. As he looked around, he noticed that part of the floor appeared disturbed. The floor was made of dirt and was covered with flat stones. "I scratched with the end of my walking stick," said Osborn, "and the more I examined, the more suspicion was created." When the men removed a few stones, they found "a human body doubled up in a heap, entirely covered up with the grave clothes."

Smith's father, Laban Smith, had been watching Osborn's men throughout. He must have been heart-struck to see his darling daughter balled up in a hole, though he carried on gamely and "rejoiced at the discovery." Professor Knight immediately claimed "he knew nothing of [the body's] being brought to the college or [of it] being there at all." Knight had Smith's face washed and her body covered with a sheet. The bereaved father, "with great difficulty," was persuaded to allow Bathsheba's body to be loaded on a wagon and then transported to West Haven for interment.

In the meantime, Yale officials ordered that all medical students remain in the building and then locked the doors in anticipation of mob violence. Hundreds of citizens soon formed. "A drum has beat," recalled Osborn, "and the streets are crowded with the besieging army preparing for the assault." Osborn had gone home after the body had been discovered and decided he would "keep at home and let the ferment have vent or subside." The mob formed at the college every night for days, stopping short of violence each time. By the end, only one student was charged, the others having surreptitiously vacated the city. That student, Ephraim Colborn, had barely escaped being tarred and feathered as he was leaving the courthouse. Luckily, he spotted the mob approaching and ran to the rear of the building, there to stay until the rioters dispersed. He was convicted by a jury of "aiding and assisting in opening the grave and removing the body of a female from the burying ground in Orange, for the purpose of dissection." The unlucky lad was finally convicted and sentenced to nine months in the county jail and fined $300.

An article in the *Connecticut Herald* a few days after the discovery of Smith's body intoned, "While we acquit the medical faculty of any knowledge of this base transaction, we freely accord to the perpetrators of the deed the general execration which it has excited. If subjects are necessary for dissection in the progress of anatomical instruction, [then] the cerements of the grave, where the manes of the loved and the lamented are placed to mingle with their kindred earth, are not to be violated with impunity, and the hand that could ruthlessly touch the hallowed spot is even more venomous than slander."

Numerous other riots occurred over the coming century, including mob actions in Zanesville, Ohio, in 1811, Cincinnati in 1839, and St. Louis in 1844. Violence broke out at the Cleveland Homeopathic College in 1852 when the remains of a young woman were discovered nearby. The female's father identified the hand of his daughter in the anatomy lab, and a mob attacked the college. Rioters broke sixty windows, wrecked the museum and library, and set fire to the building twice. The fires were extinguished each time. Michael Sappol, author of *A Traffic of Dead Bodies*, said that resurrection riots "happened at Harvard and happened at Dartmouth, and every single medical school in antebellum US had a body-snatching scandal and often a riot. In order to have a medical school you had to have a course in anatomy with dissection. No good legal source of bodies. Therefore, body snatching."

Anatomy students and, rarely, their professors were arrested, convicted, and jailed. Indeed, anyone associated with body snatching was at risk of arrest and prosecution. Take the case of a Rhode Island judge named John Dorrance. Dorrance had served as an attorney for and was a close friend of Pardon Bowen, an influential physician in Providence. Bowen taught anatomy and, at one point, was provided the body of a transient who was said to have hanged himself along a roadside on February 12, 1799. The man had been buried, snatched by Bowen's students one night, dissected over several days, and then reburied.

However, Dorrance was a staunch Federalist, and the governor of the state, Arthur Fenner, was just as staunch an anti-Federalist. Fenner had heard a rumor that Judge Dorrance had been involved in an illegal body snatching. He started telling people that Dorrance, until that time a paragon of the community, had traded a body to an anatomist in exchange for a beaver hat, worth as much as $15, or about $385 in 2025. The governor "was impulsive, profane, shrewd, and enormously popular," wrote author Benjamin Clough. "He not only told the story of the corpse and the hat to all his friends, but carried in his pocket a written copy [of the story], which he showed upon any occasion, or none, to all who would read it, especially to members of the Rhode Island legislature." He must have hated the judge; he told someone that "there are two or three damned whore's-birds whom I intend to pay for the ill treatment I have received from them, and amongst the rest, your Judge Dorrance."

In 1801, Dorrance ran for a position as judge in the court of common pleas, winning but not by as much as he thought he should have. So he sued Fenner for slander. Dorrance insisted that he had no knowledge of anyone stealing or dissecting the transient's body and that the beaver hat was a payment for services rendered to a patient. That patient, anatomist Bowen, always offered to pay Dorrance for his legal services, but because they were friends, Dorrance always refused. Bowen was finally able to persuade Dorrance to accept a hat from a local hatter. Fenner, however, continued to tell people that Dorrance knew about the body snatching and that he had indeed received the hat from Bowen in exchange for the body. Fenner's lawyer presented precisely zero evidence to the jury, but such was the governor's power and influence that the jury acquitted him anyway.

Dorrance was required to pay Fenner $10,000, plus court costs. Dorrance also later lost an election for a seat in the state's general assembly. "Although Judge Dorrance was of good character," wrote author Suzanne Shultz, "and by all accounts well-liked, it appears that the mere charge of body snatching was sufficient to cost him the election."

As the 1800s wore on, body snatching continued unabated in young America, as did rioting against it. Across the Atlantic, however, body snatching was taking an evil turn. In Scotland, two men, both named William, became infamous for practicing body snatching with a cruel twist. Rather than waiting for someone to die and be buried, William Burke, William Hare, and an accomplice named Helen MacDougal decided to do away with the waiting and the digging of all that dirt and instead execute people themselves so they could sell more pristine bodies to anatomists.

CHAPTER 13

BURKING, BONE BILLS, AND EMBALMING

WILLIAM BURKE WAS BORN IN 1792 IN COUNTY TYRONE, IRELAND. He tried his hand at a number of trades, including baking, weaving, and cobbling, but found none to his liking. He joined the militia at age nineteen as a fifer and became personal assistant to a regimental officer. He moved sometime around 1817 to Maddiston, Scotland, and found employment on the Union Canal, a thirty-two-mile-long canal then under construction. Burke was described as "a neat little man of about five feet five, well proportioned, especially in his legs and thighs—round-bodied, but narrow-chested—arms rather thin, small wrists, and a moderate-sized hand. A very active, but not a powerful man." Burke would eventually meet a "woman of disreputable life" named Helen MacDougal, and one William Hare, a man with a "ferocious and malignant disposition," according to the *Newry Telegraph*. They ended up in the same boarding house in Edinburgh, their combination making for a toxic stew.

It came to pass that on November 29, 1827, a tenant in Hare's house died, a man named Donald. His rent of £4 was coming due, so his death meant that Hare wouldn't be paid that month. Hare knew, as did most people in Edinburgh, that an anatomist named Robert Knox often purchased bodies for his students. So Hare enlisted the help of his housemate and sold Donald's body to Knox for just over £7. One of Knox's assistants told the men that "they would be glad to see them again when they had another body to dispose of." Burke and Hare conspired, then, to find suitable subjects to kill and sell to Knox. So began a year-long killing spree, during which the men killed seventeen unfortunates. MacDougal, who by that time had married Hare, fled the city and escaped persecution, as did William Hare. The ringleader, Burke, was hanged for his crimes. His surname turned into a verb: to burke, meaning to kill a person and have their body dissected.

The murders, along with the British public's deep concern about body snatching and the abundantly clear need for more cadavers for medical students, prompted Parliament to pass "An Act for Regulating Schools of Anatomy" in 1832. The bill stated clearly that medical students needed to learn anatomy, that they needed to have access to cadavers to learn best, that an insufficient supply of cadavers impeded their knowledge and, thus, the future healthcare of the Crown's citizens, and that "great and grievous crimes have been committed" in search of that supply. The act tightened restrictions on anatomy education and, more importantly, authorized the lawful use of cadavers unless the individual had directed "either in writing at any time during his life or verbally in the presence of two or more witnesses during the illness whereof he died."

Even though the wording of the act left unclaimed bodies to be used for dissection at will, the act still led to a remarkable decrease in body snatching almost as soon as it passed. America was making similar, but ineffective, inroads into the issue. There, body snatching and its attendant issues were being dealt with by individual states, which tried to walk a delicate line, aiming to reduce the incidence of body snatching while also allowing the growing number of medical schools to obtain cadavers legally. Only four medical colleges existed in America by the end of the eighteenth century: the College of Philadelphia, Columbia College in New York, Harvard Medical College in Cambridge, Massachusetts, and Dartmouth Medical College in Hanover, New Hampshire. The number exploded during the first part of the next century. Nineteen additional medical schools opened up by 1840 and another forty by 1860.

To meet the needs of the new schools, states began passing anatomy acts, the first of which was passed by New York in 1789. The bill, "An Act to Prevent the Odious Practice of Digging Up and Removing for the Purpose of Dissection, Dead Bodies Interred in Cemeteries or Burial Places," banned body snatching completely. It also ordered that anyone convicted of it "shall be adjudged to stand in the pillory or to suffer other corporal punishment, not extending to life or limb." The individual would furthermore be required to pay a fine. In addition, so that "science may not in this respect be injured by preventing the dissection of proper subjects," the act allowed judges to sentence individuals executed for arson, burglary, or murder to be "delivered to a surgeon for dissection."

New Hampshire enacted in 1796 a similar anatomy act, or "bone bill," as those early anatomy acts were known, that outlawed body snatching. Perpetrators could be fined up to $1,000 (about $25,000 today), sent to prison for up to a year, or tied to a post and whipped not more than thirty-nine times. Massachusetts passed a slightly more progressive act on April 1, 1834, that banned body snatching and allowed the body of executed criminals to be dissected. However, it also permitted dissection of unclaimed bodies. Anatomists receiving a body were to pay a "good and sufficient bond" to ensure that the body would be used "only for the promotion of anatomical science." Eventually, most states passed some form of bone bill.

None of the bills provided for enforcement, however. "The vast majority of illegal disinterments were not discovered," said medical historian Frederick Waite, "and when detected the offenders were rarely apprehended. Arrests and indictments were few, and convictions yet fewer." Body snatching continued unabated as the number of medical schools increased. More medical schools meant more students, which, in turn, meant more bodies needed for anatomy education. While the need for cadavers climbed, the number of bodies available for dissection stayed essentially the same unless some widespread infection caused a higher than normal number of deaths.

Medical schools in states with lower populations began looking to higher population areas for cadavers. New York proved a most prolific source. Schools in New England purchased many bodies from New York "at a great expense of money and great hazard of being discovered." The Medical College of Georgia hired freelance resurrection men to import bodies from New York. Between 1848 and 1852, sixty-four bodies were purchased in New York and transported to Augusta in old whiskey barrels. Faculty at the school decided in 1852 to bring resurrection duties in-house and purchased for $700 a tall, strapping man named Grandison Harris. His duties at the college were officially listed as janitor and porter, but his actual duty was to procure bodies for the anatomy lab.

Photo of the Georgia Medical College (now part of Augusta University) class of 1880. Grandison Harris, known as Resurrection Man, stands in the top row, far right, waving. COURTESY OF HISTORICAL COLLECTIONS AND ARCHIVES, ROBERT B. GREENBLATT, MD LIBRARY, AUGUSTA UNIVERSITY.

Late at night, Harris would sneak into the Cedar Grove Cemetery, a few blocks south of the college, and fetch recently interred bodies. The cemetery housed the bodies of Black citizens only, most of whom had been buried in wood coffins or simply wrapped in cloth and interred in unmarked graves. Harris proved untouchable as a resurrectionist. He couldn't be arrested for his trade because he was a slave, and because his employers were some of the most eminent men in town, they too were safe from prosecution. Of course, if someone raised a ruckus, Harris would have been conveniently put forth as the culprit. After Harris died in 1911, at around age ninety, he was buried in the same graveyard he once plundered.

Where bone bills couldn't stop body snatching, modern embalming practices finally did. The first significant development in embalming techniques in America came during the Civil War. Death prior to the war was largely a local phenomenon. People died where they lived. In the war, though, soldiers traveled great distances from their homes to the front lines. Those killed in action needed to be transported back to their homes for burial. An embalming technique had been developed by a French chemist and involved infusing a combination of mercury, turpentine, creosote, arsenic, and alcohol into the body.

The first test of the new technique came just six weeks after the start of the war, on May 24, 1861, after a soldier in Alexandria, Virginia, had been killed in a skirmish with a rabid secessionist. The soldier, Colonel Elmer E. Ellsworth, had known President Abraham Lincoln when he was just a gangly, odd-looking lawyer from Springfield, Illinois. When Ellsworth was killed, Lincoln was gutted. Lincoln said to those gathered near him when he heard the news, "I will make no apology, gentlemen, for my weakness, but I knew poor Ellsworth well and held him in great regard. The event was so unexpected and the recital so touching that it quite unmanned me."

Thomas Holmes, an ambitious embalmer from New York, had been using the new technique, so he rushed to Washington and offered to embalm Ellsworth. Lincoln accepted. Holmes then embalmed Ellsworth by making a small incision in the arm and inserting a needle attached to tube connected to a pump. He turned the pump on and pushed about two gallons of fluid through the tube. The embalming fluid pushed blood through the body and out again through a tube attached to a vein. The fluid preserved Ellsworth's body long enough for it to lie in state at the White House for several days before being buried in Mechanicsville, New York, Ellsworth's hometown.

At Lincoln's direction, Holmes set up a military team of "embalmer surgeons" and supervised them for a year before resigning his commission. He had realized how much money could be made embalming and opened his own practice. Holmes, who would earn the nickname the "Father of Modern Embalming," embalmed an estimated 4,000 Union dead, including eight generals, many at a fee of $100 per body. Approximately 40,000 casualties were embalmed during the war, out of 698,000 killed, North and South combined. When Lincoln was assassinated on April 15, 1865, just after the war ended, his body was embalmed for the great man's cross-country train ride home.

Embalming practices changed completely after the discovery in 1867 of a chemical called formaldehyde. A German chemist named August Wilhelm von Hofmann identified a colorless gas with a noxious, stinging odor, a gas scientists now believe might have played a role in providing early Earth with carbon and other essential elements. The human body constantly produces formaldehyde through normal metabolic processes, and the gas is encountered in tobacco smoke, automobile exhaust, and forest fires. Formaldehyde didn't find any particular use after it was identified until a German physician, Ferdinand Blum, discovered quite by accident that the gas could be dissolved in water and used as an embalming agent.

Blum noticed that after working with a solution of formaldehyde, the skin on his fingers became hardened, not unlike the hardening that occurs when working with alcohol. Histologists at the time would immerse a tissue sample in alcohol and then cut the now-stiffened sample into extremely thin slices for viewing under a microscope. Blum tried immersing tissues in a formaldehyde solution and found it worked just as well as alcohol to cause stiffening. Better yet, the solution didn't distort tissues the way alcohol did. Today, scientists know that formaldehyde performs three critical functions. First, it sanitizes organs and tissues, thereby cutting the rate of decomposition to nearly zero. Second, it removes water from tissues. The human body is comprised of 50 to 70 percent water. Removing that water helps preserve tissues. Third, the chemical reacts with proteins in cells to fix, or lock in, the shape of the tissues.

A solution of formaldehyde and water, or formalin, eliminated the need for cadavers to be dissected immediately, before decomposition set in. "Cadavers obtained from legal sources could be accumulated over the period of time when classes were not in session," said author Susan Shultz, "and used as the need arose. Dissections could be performed with more attention to detail, in a less hasty fashion, and in a more complete manner." Over time, medical schools came to embalm all of their dissection subjects. Sometimes schools used funeral homes to embalm bodies, and other times, staff or students at a school embalmed bodies themselves. As the number of medical schools using formalin grew, the incidence of body snatching dropped. By the early 1900s, the practice had all but disappeared.

Concomitant with the widespread use of formalin came other advances in medical education and, most important, changes in laws regarding the use of cadavers. The organized arm of medicine, the American Medical Association, formed in 1847, pressured lawmakers to pass laws that recognized the bond between anatomy education and general medical education and that one could not exist without the other. Medical education became more refined and comprehensive, and with that came renewed respect for doctors throughout society.

By 1954, medical schools existed in thirty-nine states, all of which had passed laws allowing the use of the unclaimed dead, defined as "those bodies not assumed by friends or relatives for burial or those dead who would be interred at public expense." Half of those states, though, mandated a twenty-four- to seventy-two-hour waiting period before a body could be delivered to a college for dissection, to allow time to find family members. In almost all states, with medical schools or without, laws remained on the books that prohibited transporting a body across state lines. Those laws finally began to change in 1967 when a young cardiac surgeon from Cape Town, South Africa, performed a historic operation on a dying man.

Christiaan Barnard was a tall, thin, handsome man with large ears and brown hair, neatly parted on the left. Charismatic, confident, and sometimes arrogant, Barnard had performed many open-heart operations at the University of Minnesota Hospital, where he trained, and later took those skills to Groote Schuur Hospital in Cape Town, South Africa. It was late in 1967 when Barnard walked into the hospital room of an extremely ill fifty-three-year-old diabetic named Louis Washkansky. Barnard proposed to the dying man that he become the first recipient of a transplanted heart. Washkansky agreed immediately. "For a dying man it is not a difficult decision," Barnard said later, "because he knows he is at the end. If a lion chases you to the bank of a river filled with crocodiles, you will leap into the water convinced you have a chance to swim to the other side. But you would not accept such odds if there were no lion."

Washkansky lived only eighteen days after receiving a new heart, dying not from the transplant but from pneumonia. His transplant followed other historic organ transplants. The first human-to-human kidney transplant took place in 1954, the first liver transplant in May 1963, and the first lung transplant just a month later. Medicine was delving into new worlds, and with Barnard's success, the nation needed new laws to adapt.

The legal profession responded. In 1892, the American Bar Association supported the creation of the Uniform Law Commission to help states provide greater unanimity of laws. In the mid-1960s, the

Commission began to address the medical use of human tissues. The group had been working through drafts of a document for two years before Barnard's surgery and approved the final document less than a year later, on August 6, 1968. The document, called the Uniform Anatomical Gift Act, provided a blueprint for states to create and adopt as their own. All fifty states and the District of Columbia approved their own versions of the act, most doing so within four years.

The act established that the body was that individual's property and that a donor's wishes legally superseded those of their next of kin. If a deceased person never made a choice, either for or against donation, the family could then decide. The act opened the doors for medical schools to obtain legally donated bodies for education. Over the ensuing decades, the general public became steadily more open to whole-body donations, particularly after medical professionals began donating their own bodies. Today, many medical schools operate a willed-body program for individuals and families interested in registering for body donation. Individuals wishing to donate their body register with the program. When the donor passes away, a family member contacts the program, which assigns someone to oversee the body from the time it is collected to the time the cremated remains are returned to the family. "We're stewards of your body," said Margaret McNulty, an educational vice chair at Indiana University's medical school, "and we take that stewardship very seriously. We pass that feeling on to our students as well."

Unfortunately, as whole-body donations and willed-body programs increased, so too did a new form of body snatching emerge: the sometimes unethical use of human body parts. Although organ donation is tightly controlled, the donation of a cadaver or its body parts is essentially unregulated. Today, a number of body parts can be procured legally, transported legally, and transplanted legally. Tendons and ligaments taken from a cadaver can be used to repair joint damage from sports and other injuries. Bones taken from cadavers can be used to replace cancerous bones, provide support in spinal surgeries, or increase bone mass for dental implants. Collagen taken from cadavers can be injected into the face, lips, and buttocks for cosmetic or plastic surgical repairs. Skin taken from cadavers can be used to repair burns or wounds, and veins can be used in heart surgeries. Entire shoulders, feet, hands, heads, and other body parts can be used by surgeons practicing new procedures, paramedics learning how to intubate someone, and even engineers making new kinds of surgical instruments and medical devices.

A whole industry has developed to meet the demand for body parts, often referred to by a softer, less controversial term: tissues. Besides willed-body programs, a number of independent non-transplant tissue banks, commonly referred to as body brokers, have developed. Body brokers obtain most cadavers from funeral homes, hospices, and hospitals, and sell the entire body or separate body parts—shoulders, legs, feet, heads, and so forth—to universities, pharmaceutical companies, for-profit anatomy education centers, medical device manufacturers, and the like. An investigative report by Reuters in 2017 found that body brokers usually section cadavers into six parts, pricing each part separately. A torso with both legs attached sold at the time of the report for an average of $3,575. The fee for a head averaged $500, a single foot $350, and an entire spine $300. The report found that annual revenue for a single body broker in Tennessee rose from just under $50,000 in 2009 to more than $1 million in 2016.

Body brokers tend to operate below the surface of society and lack anywhere near the oversight given to organs donated for transplant. Although selling an organ for transplant is illegal in the United States, there are no laws blocking the sale of human body parts. Tanya Marsh, a law professor at Wake Forest University, said that tissue-selling "doesn't fit within the regulatory structure of the healthcare industry. It doesn't fit within the regulatory structure of the funeral industry. And it just doesn't really seem like anybody is watching the folks who are engaged in this." Famed bioethicist Arthur Caplan concurred, saying, "It's a wild, wild West out there in tissue land, with few sheriffs and a lot of shady characters meeting in the back rooms."

One of those shady characters was a fraudster named Michael Mastromarino, who made millions from selling bones, arms, legs, torsos, and other tissues to legitimate medical companies. One of the many bodies dismembered and sold under Mastromarino's direction was the body of Alistair Cooke, the congenial, silky-voiced host of the PBS series *Masterpiece Theatre* and the host of *Letter from America*, a fifteen-minute weekly radio broadcast on BBC that ran for an astounding fifty-eight years.

Cooke died at ninety-five from metastatic lung cancer, and at his family's request, his body was sent for cremation to New York Mortuary, a funeral home in Harlem, New York, one of several used by Mastromarino. The mortician called Mastromarino and told him that Cooke's body was on his embalming table, asking what he should do. Mastromarino said simply, "Proceed." With that, Lee Cruceta, a nurse and tissue bank specialist working with Mastromarino, deboned Cooke's arms and legs, cut out his heart, carved out the heart valves, and then cut out or off everything else of use. The remainder of Cooke's remains were cremated and sent to Cooke's daughter, Susan Kittredge.

Kittredge knew nothing about what had happened to her father's body, only that it was supposed to have been cremated and that she would be sent the remains. The families of Mastromarino's other victims likewise knew nothing about any tissues removed from their loved ones. Mastromarino forged signatures on consent forms and nearly always falsified patient records, changing not just the decedent's name but often the date of birth and medical history as well. Tissues from younger, healthier bodies were more likely to be accepted by processing companies. Cooke's date of birth, November 20, 1908, was changed to one in 1918, making him ten years younger, and his cause of death was changed from metastatic cancer to heart attack, so as not to seem quite so sick at the time of his death.

Numerous other modern-day body snatching scandals have occurred. In one, a disgraced chiropractor operated an illicit body brokerage located in a Las Vegas strip mall between a psychic's shop and a tattoo parlor. The chiropractor, Obteen Nassiri, purchased whole or dismembered bodies from funeral homes and then sold or leased them. He sold, for example, six pairs of legs to a client in Chicago for training in implanting medical devices, such as artificial hips and knees. He sold five sets of arms to a client in Pennsylvania for a refresher course in shoulder surgery, twelve whole bodies to the US military for dissection training at a medical school in Bethesda, Maryland, a half-dozen torsos to a Massachusetts client for training, and two more to a spinal surgery conference in Hawaii.

Dozens of Nassiri's cadavers came from the willed-body program at the University of North Texas Health Science Center in Fort Worth, which had been collecting unclaimed bodies and leasing them to research companies and the military. A survey in 2023 found that eight of fourteen medical schools in Texas used unclaimed bodies for dissection and that the number of unclaimed bodies received by those schools had been increasing steadily each year, from just over 2 percent in 2017 to over 14 percent 2021. When a person dies intestate, whether in a hospital, nursing home, their home, or on the street, morticians and medical examiners try to find family members to determine the method of the body's disposal. When no relatives can be found, bodies can languish in a morgue. If the morgue becomes overwhelmed with bodies, if they lack the funds or personnel to locate relatives, or if they just don't bother, the coroner may elect to send the body to a medical school or, knowingly or unwittingly, to a disreputable body broker like Nassiri. Missing from all of those transactions was any form of consent. The families never consented to have their loved one's body split up and sent to other states, nor did they consent to have the body displayed in a conference room at a trendy resort.

Victor Carl Honey had served in the US Army for ten years. Later in his life, he suffered from mental illness, was often homeless, and died intestate in Dallas in 2022. The county's medical examiner reported that the office tried to find Honey's relatives but couldn't. It turned out that Honey had a son with the same name, and he too lived in Dallas, as did his ex-wife, Kimberly Patman. She told NBC News that Honey "never wanted to be an organ donor. We talked about it." She also said that she had never been contacted by anyone about Honey's death. "I don't believe they tried to find us," she said. "You can find people. NBC found me in a day. You can find people."

Honey's body was sent to the University of North Texas, where it was dismembered and the pieces sold. A medical device maker in Sweden purchased the right leg for $341. A medical education company in Pittsburgh purchased his torso for $900. After NBC reported on Honey, the university apologized to the family, suspended its body program, and fired the program's leadership, but by then, the damage was done.

Illegally sourced body parts have also found their way into the oddities trade. Oddities are unusual and often macabre items, such as crystals, bug art, taxidermy specimens, pagan items, and animal and human bones. Oddities markets are held throughout the nation and resemble country flea markets. Many oddities dealers today use social media to advertise their wares, which is how an Arkansas woman named Candace Scott found a heavily tattooed Pennsylvania man with metal spikes embedded in his head. Jeremy Pauley, an oddities dealer near Scranton, described himself as a "curator to historic remains and artifacts." Scott came across a Facebook post by Pauley, and the two soon set up a business arrangement.

Scott worked at a funeral home in Little Rock that often received bodies from the Medical Sciences Anatomical Gift Program at the University of Arkansas. She collected body parts and then, over a nine-month period, sold to Jeremy Pauley twenty-four boxes of human remains, including several lungs, hearts, breasts, and testicles, as well as a brain, a skull, an ear, an arm, a belly button, and two whole, stillborn infants. Pauley used the payment application PayPal to send Scott $10,625 for all boxes. Pauley turned around and sold one of the infants to a Minnesota man named Matthew Lampi, who later was sentenced to fifteen months in prison for obtaining and transporting human remains across state lines. Pauley pleaded guilty to state charges of abuse to a corpse and to federal charges of conspiracy and interstate transportation of stolen property.

Not even one of the oldest and most prestigious medical schools in the country escaped the Pauley scandal. In June 2023, a morgue employee at Harvard Medical School was accused by authorities in Pennsylvania of the theft and transportation of human remains across state lines. The employee, Cedric Lodge, worked for the school's Anatomical Gift Program from 2018 to his termination in August 2022. He pled guilty in May 2025 to interstate transport of stolen human remains.

Anatomy professors, bioethicists, and law enforcement officers alike agree that the laws governing the use of human tissues need to adapt to the times. Paul Micah Johnson, special agent for the FBI, investigated the body broker business in Detroit for more than ten years. He understood the need for donor bodies for education and research but also understood perhaps better than most what happened when those bodies weren't being used as intended. "We want our surgeons to practice before they operate on us," said Johnson, "and for the medical device companies that are developing all of these prosthetics and implants to have real human body tissue. That's what this industry is for, and that's what it does well." The problem, he added, is that "if you have a Sawzall and a garage, in a lot of states you can go into the business."

Angela McArthur, director of the willed-body program at the University of Minnesota Medical School, agreed and said that "the current state of affairs is a free-for-all. We are seeing similar problems to what we saw with grave-robbers centuries ago." Thomas Champney, an anatomy professor at the University of Miami School of Medicine, said that body broker scandals are "rare," but admitted that no data exist to say for sure one way or the other. "Right now there are things going on that we don't know about," said Champney. "There is no oversight. There is no national registry for body donation, so we just don't have the data we need. I've been pushing for federal regulation of body donation programs for fifteen years."

Champney said that until suitable legislation is passed at the federal level, consumers need to exercise great diligence in selecting a donor program, particularly one of the many non-transplant tissue banks. McArthur urged anyone considering donating a body to a broker to make sure the bank is accredited by the American Association of Tissue Banks (AATB). "The AATB has very robust standards," said McArthur, "and conducts regular inspections of donor facilities." She added that consumers should look at the length of time any particular donor program has been operating. "For-profit companies usually have a short history," she said, "whereas most university programs have been around decades." University programs are also typically subjected to internal audits, and those reports are available to the public. "There is more transparency in those kinds of programs than for-profit companies."

Most of the more public of body broker scandals over the recent past, however, have been centered on college-based programs, making the potential donor's decision about where to donate their body difficult. Champney compared making that decision to buying a house. "It's a major life decision," said Champney, "that needs to be considered carefully. Potential donors should always investigate the program as much as possible, including visiting the facility and making sure they get answers to all of their questions." Funeral home directors, officials at university-based willed-body programs, and especially directors of non-transplant tissue banks should answer questions honestly, clearly, and completely and be willing to show prospective clients their facilities.

Donors should ask how long a loved one's body will be used (generally one to three years); what costs would be charged for transportation, death certificates, and cremation; what all of the possible uses of the body would be; whether the body will be transferred to other institutions; whether the body will be photographed or used in public demonstrations; and what the final disposition of the remains will be.

Experts say donors should also be aware of receiving inaccurate or misleading information—from body brokers or willed-body programs alike. Among the most common pieces of misinformation provided to clients is that donating one's body can lead to cures for diseases. FBI agent Johnson called such information "garbage. That just doesn't happen." The majority of donated bodies, he said, are cut into pieces and sent to medical device manufacturers and facilities that use them for training healthcare professionals. He explained that some donors are sent to automotive manufacturers for crash testing, and others end up in military research facilities for testing the effects of bombs or other weapons on the body. Virtually none are used to research disease cures.

Regardless of what happens with a whole body or its parts, someone is paying. Funeral homes are being paid for sending bodies to donation companies, which are, in turn, paid for body parts sold to medical device manufacturers and educational centers. "Everybody makes money in this," said Johnson, "except the donor."

Decisions about donating a body come down to what a person or their family want from the donation. Many individuals don't care what happens to them after death, they just want to help others. Marie Holmes's father, a neurosurgeon, arranged for his body to be donated to Oregon Health & Science University, where he taught medicine. "My father has been gone for over a decade," wrote Holmes, "and the ache I feel, that he is not here to see his grandchildren grow up, can expand or contract, depending on the day. The students he taught, both in person and with the gift of his body, are out in the world practicing medicine. One of them may now be gently palpating a patient's abdomen, imagining the shape of my father's organs."

Others might not feel comfortable with certain elements of the body-donation process and want to limit use of this part or that. "If someone wants their body used only for dissection," Johnson explained, "they need to ask the donation program what's going to happen to their body and whether they can put restrictions on it."

Restrictions might include blocking use of the body by other countries. The University of Melbourne, for instance, sometimes imports bodies or body parts from the United States. Jackie Dent, author of *The Great Dead Body Teachers*, explained that "sometimes fifteen knees are needed for a surgical workshop, and it is logistically easier to get them from a US body broker."

Other possible restrictions include no military use, no sending of the body across state lines, no use of certain body parts, or no placement of a body part in a museum. "Transparency is the main concern," said Johnson. "Potential donors should have a very direct conversation with officials of the donation program." Unfortunately, Richard and Angie Saunders seem not to have had that kind of conversation.

The Saunders couldn't afford to bury their son, Cody, who had been born with kidney and heart disorders. He underwent sixty-six operations in his short life and died in 2016 at age twenty-six. Cody's parents decided to donate their son's body to a nonprofit company in Elizabethton, Tennessee, called Restore Life. "I couldn't afford nothin' else," said Richard Saunders. Whether through miscommunication, lack of transparency, or a general misunderstanding of body donation, the Saunders believed that the company would remove some skin from Cody, nothing more, and then cremate him and return his ashes to them.

Instead, the company divided Cody's body into parts, including the spine, and sold it for $300 to an undercover journalist for Reuters. The University of Minnesota's Angela McArthur later examined Cody's remains for Reuters. She determined that Restore Life had provided insufficient medical information to the buyer and that the paperwork accompanying the body was careless and lacked the standards required by her university. "I haven't seen anything this egregious before," she explained. "I worry about the future of body donation and public trust in body donation when we have situations like this."

CHAPTER 14
ULYSSES, SIR, AND CELEBRATING DONORS

THEY CALLED HIM ULYSSES. HE HAD DIED IN HIS SEVENTIES FROM cirrhosis of the liver. His body had been donated to Slippery Rock University in Pennsylvania, where Dawn Lynch was studying to become a physical therapist. Lynch remembers vividly what the man's liver looked and felt like as a result of his long-term alcoholism. "I'll never forget it," said Lynch. "What alcohol did to this man was unbelievable. A normal liver has a maroon color. It feels like a sponge, with a little softness. But his was rock hard. It was pink, and it had a scalloped edge, like a shell you might find at the beach." To Lynch and her fellow students, Ulysses was not a corpse, nor a cadaver, nor even a dead body. He was a donor, the term used for individuals whose bodies have been bequeathed to an anatomy program for use as a learning tool.

The donor used by Howard Chang when he was a medical student at Johns Hopkins University was known simply as Sir. He was seventy-nine years old when he died from multiple myeloma, a blood cancer. "The evidence of the disease startled us," recalled Chang. "He had numerous holes on his skull, and a surgically inserted metal plate to stabilize what we presumed were damages to his spinal column from the cancer cells. It staggers me to think that we had explored every square inch of Sir's internal and external body—something no one had ever done—and yet knew nothing about his personality, character, or aspirations, the things that defined him as a person."

The first time students see their donor can be rather unsettling. "The scariest part of anatomy," said one student, "was walking into the lab the first day. Seeing the outlines of familiar body parts we see on living people every day, suddenly not living, not moving, and under thick brown tarps, was extremely disturbing," Removing the cover, though, gave the student some peace. "Once our cadaver was uncovered," she said, "I immediately felt better. I was able to see and appreciate the gift that this person freely gave to us. Knowing that she wanted us to learn with her made all my fears disappear."

After the initial incision into the donor—perhaps down the spine, perhaps along an arm or leg—some students find that the donor becomes somewhat less human and more an intricate, inanimate puzzle. "Over time," said one student, "I got used to cutting a human body and somehow began to forget that I was working on a human. We had covered her face and were working on her limb muscles for a while. When we uncovered her face to dissect her head toward the end of anatomy, I was once again reminded of her humanity." Pauline Chen, author of *Final Exam: A Surgeon's Reflections on Mortality*, explained the student's essential need to separate the body from the human who once inhabited it. "Medical students learn to deny their own feelings," wrote Chen, "depersonalizing the dissection experience and objectifying their cadaver. They strip away the cadaver's humanity, and soon enough they are not dissecting another human being but 'the leg' or 'the arm.'"

Sindhu Ragunathan, a first-year medical student at Saint Louis University, feared that spending so much time under "blinding white lights and the overwhelming smell of formaldehyde" in a dissection lab would desensitize her to the donor's humanity. She found just the opposite. "In learning about muscles, ligaments, tendons of the human body," said Ragunathan, "we learn that even the smallest structures can hold the deepest meanings. Freckles spread across shoulders and cheeks that speak of a life spent in the sun. The anatomy lab cannot possibly feel cold and sterile with so many signs of life in it."

Students during dissection class must confront their own fears about death. "Aspiring physicians face death directly in the form of the cadaver," said Chen, "and then they tear it apart." Like Chen, students will cut into the donor to identify arteries and veins, nerves and lymph nodes, muscles, ligaments, and tendons. They will excise the stomach, the heart, the liver and intestines, and then hold each in their hands, sensing its heft, density, and shape, looking for irregularities. "Each detail of the cadaver," said Chen, "every bone, nerve, blood vessel, and muscle passes from the world of the unknown into the realm of the familiar. Every cavity is probed, every groove explored, and every crevice pulled apart. In knowing the cadaver in such intimate detail, we believe that we are acquiring the knowledge to overcome death."

At times, other learners besides healthcare students arrive in a dissection classroom to study donor anatomy. Madeleine Spencer was one such learner. Spencer is an artist in London who gained access to a dissection lab while working on an anatomy textbook. Although she had been studying the human body for many years, she experienced the same kind of depersonalization that medical students go through. "I found I could not effectively work if I maintained too close a proximity to the sense of the cadaver as a person," she said. "Other times, as I worked, I found myself being sure to treat the body gently and to turn him slowly and kindly. They were never just objects to me. If I were to disconnect and never again engage with their humanity, I would do a grave disservice to their generosity."

The essence of a donor might be long gone by the time they've been wheeled into a dissection classroom and placed on a steel table, but the education they provide and the emotions they evoke last a lifetime. Students gather around the table, suppress the humanity of the donor, and immerse themselves in the learning of anatomy. They ask questions of each other about this body part or that, hoping someone in the group has the answer.

"Is that the cystic duct?" someone might ask.

"Yes, I think so. Wait. No, that's the cystic artery. See? There's Calot's lymph node."

"So, then, this one must be the common bile duct. It's not as thick as I thought it would be."

Discussions like that abound as students work their way quite literally through a donor's body. Like photographs of the Grand Canyon, the anatomy they've seen in books can't compare to what they see in the lab. "I had studied anatomy books for years," said Spencer, "from hundred-year-old volumes with hand-drawn plates to modern photographic editions complete with axial MRI scans. None of them captured the nuances and beauty of the human body as seen in person. One particular feature leapt out at me: The tendinous bands in the body, those flat bands of tendon that extend from the belly of the muscle to the ends, have a shimmering quality not unlike mother of pearl and a pronounced iridescence that's nearly impossible to photograph."

Technological advances are getting closer to that ideal, however. Medical school anatomy courses today take advantage of technologies that use data gleaned from a variety of sources. The well-known Anatomage table is one such technology. The Anatomage table resembles a body-sized pool table, except that instead of felt, there is a touch-sensitive display of layer after layer of the human body. The table uses highly detailed images of head-to-toe cross-sections of frozen cadavers donated for research, many of them from the Visible Human Project. The images were put through a digital transformation process that, in a way, disassembled and then reassembled separate images into a nearly three-dimensional body. Students can manipulate the display of any section of the body and scale it down to five millimeters. The product can also show motions of the heart and eyes, certain muscle movements, and even the delivery of an infant.

Among the more exciting technologies are those that use virtual reality components, augmented reality components, or a combination. The term virtual reality refers to a digital simulation of a

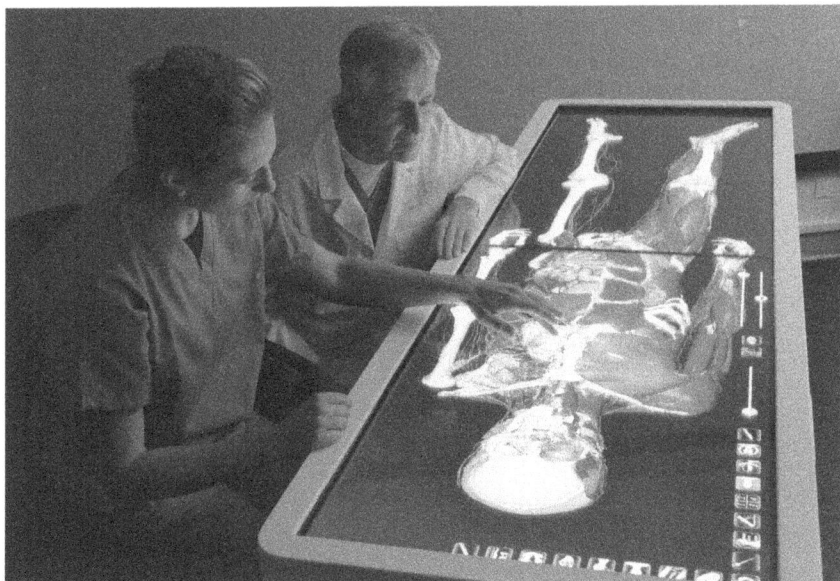

Two healthcare professionals working at an Anatomage table. COURTESY OF ANATO-MAGE, INC.

three-dimensional, completely artificial environment and is typically shown on a display screen built into a headset. After a period of adjustment, the student's brain learns to accept the artificial environment as real. Virtual reality games can be extremely lifelike and provoke in the user the same kinds of reactions they would experience in real life. Bumping into a virtual wall, for example, can make the viewer rear back as if the wall was real.

Virtual reality games are plentiful today, and the number of virtual reality simulations for anatomy is growing. HoloAnatomy is one such innovation. The product is a software application that runs on HoloLens, a visual-display headset from Microsoft. The software allows students to observe a digital body in great detail and interact with that body within the confines of the student's physical space. A student can, for instance, hold a digital scalpel and dissect a holographic shoulder of a patient lying on a holographic operating table.

Augmented reality takes virtual reality one step further by super-imposing computer-generated images on a user's view of the real world, in real time. For instance, with an augmented reality headset a student could walk up to a real person acting as a patient, place a digital stethoscope on the person's chest, and hear normal, abnormal, or no breath sounds, depending on the simulation being used. Anatomy applications for mobile phones have been developed as well, though their effectiveness as a learning tool tend to be limited by the amount of storage in the phone and the speed of the internet connection. Simulation-based products are also being developed using artificial intelligence tools, though that technology is still too new to evaluate its effectiveness in anatomy education.

Virtual and augmented reality technologies possess many advantages over cadaveric dissection, including the ability to be used for remote learning. Many more students over the course of a year can use virtual patients than they could a human donor. The technologies can also put images of normal anatomy side by side with those of abnormal anatomy, a comparison more difficult and cumbersome to make in a dissection lab. Students also tend to feel more comfortable using a virtual or augmented reality system than they do standing by a donor body in a lab, holding a scalpel in their hand. Students studying, say, the path of an artery from the heart to the foot can easily isolate the artery in a digital environment, a much more laborious process to complete on a donor. Students working on donors in a dissection lab can see only the kinds of pathology the donors themselves had during life. Digital donors, on the other hand, can be used to study a much wider variety of disorders.

All of those advantages, however, come with higher costs to medical schools and anatomy students. At this writing, a single Anatomage table can cost more than $50,000, depending on its features. In 2024, one US university purchased a four-year license for a bundle of HoloAnatomy software, devices extra, at a cost of $100,000. At some colleges, the students are required to purchase their own devices, which can cost several hundred dollars apiece, a price many students find beyond their means.

Cost aside, several studies suggest that virtual and augmented reality technologies can be enormously helpful for students trying to learn the intricacies of the human body. Whether they could ever replace cadaver dissection is another question. One study during the COVID-19 pandemic determined that traditional dissection provides "an irreplaceable tool for familiarizing [students] with fine movements in a stress-free environment." The study also found that traditional dissection nourishes the students' ability to empathize with patients, improves their collaborative skills, and enhances their professionalism.

The impact of a human donor on students can hardly be overstated. Every donor was a person who gave the last physical evidence of their life on Earth to others, even if those like young Cody Saunders, veteran Victor Carl Honey, and the legendary Alistair Cooke might not have planned it that way. Zelda Blair wrote about how much her donor meant to her when she was a student at the University of Minnesota Medical School. "We're told the donor is your first patient," she said, "and that's true. I think of this in terms of the way someone's name changes its sound when you love them. I don't know the donor's name, but my acquaintance with him is as intimate as anything I've ever experienced. It's something closer to pure humanity. I will never forget this experience. It has taught me a lot more than anatomical science."

Most anatomy programs that use human donors hold reflection ceremonies at the end of each course. Students and faculty invite friends and family members of people whose bodies have been donated to the school. The families are not necessarily those of that year's students, however; the identity of most donors remains unknown. During the ceremony, students address the families and explain how important their particular donor was to them.

A graduate student in anatomy at Ohio State University's medical program spoke at the program's memorial service in August

2020, saying, "These future doctors, dentists, and educators you see sitting in front of you will have numerous encounters with patients and students throughout their careers, and many lives will be saved and changed for the better. Without the generous and selfless donation that your loved one has made, none of this would be possible. We hope you realize what a lasting impact this gift will have on our society as a whole, and we thank you from the bottom of our hearts for your gift and support."

University of Houston student Stephanie Huffman echoed those comments, saying, "We have a huge responsibility to honor their sacrifice, their gift. The best way we as medical students can do that and respect what they sacrificed is to work hard and gain as much knowledge as we can from them, so we can one day make a difference in the lives of our patients." Thousands of students around the world have likewise expressed their intense and heartfelt gratitude to the people who donated their body and the friends and families who loved them.

Perhaps no one has summed up those feelings as eloquently as Samantha Cooper. Cooper was a member of the 2025 class of the Pennsylvania College of Osteopathic Medicine when she completed her anatomy course. The college held a "Celebration of Remembrance" on May 3, 2022, on the campus of the University of Pennsylvania. A vase filled with white roses, one rose for each donor, was placed outside the anatomy lab to remind anyone passing by of "the sacrifice the donors have made to advance scientific research and aid in educating future healthcare professionals."

Cooper stood before her audience and spoke not only to her donor, but also to donors and their families everywhere. She told the group,

I wish I could express the significance of what your donation means to me and my fellow peers. I have held your heart in my hand, but I do not know for whom it once beat. I have held your brain, but I do not know the memories within the folds. To learn about the hands that could have held your first child, the feet that helped you take your first steps, and the body that has carried through your life. I have seen every bone, muscle, organ, nerve, artery, and vein that makes you, you; and as you taught me every name of every part of you, your own name was a mystery. I do not know how I could ever repay such an immense gift, but I thank you for all you have done for me and my peers, and I hope that I can carry on the lessons of the beautiful life you lived.

The measure of a life, after all, is not its duration, but its donation.
—CORRIE TEN BOOM

ACKNOWLEDGMENTS

My writing space for this entire book has been an eight-by-ten-foot room that once housed my late mother-in-law's roll-top desk, computer table, and two bookcases, each of them piled high with years' worth of bills, old magazines, and ancient birthday cards. The room today still holds a desk, a table, and two bookcases, but it seems like home to me. On the walls are items that bring me comfort. There's a black frame with a single fork glued to the center of a white background. The fork, with a raffia bow on the handle, is engraved, "I'm done, 2016," a classic line from *Seinfeld*—"Stick a fork in me, Jerry, I'm done"—and given to me by a dear work colleague on my retirement.

There's a bright red golf towel from The Broadmoor, a fabulous resort in Colorado Springs with three superb golf courses, a reminder of the many outings to world-class courses I was fortunate enough to play during my working years. (Thank you again, Rob.) There's a watercolor of the head of a bald eagle, painted by my freakishly talented wife. There's a framed print of a map of Donora, Pennsylvania, given me by a friend I met when writing my first trade nonfiction book. There's a huge print of our former hometown, Doylestown, Pennsylvania, and on a shelf directly in front of me is a hand puppet of Mr. Gopher, the scene-stealing character in one of my favorite movies, *Caddyshack*. He's a reminder to keep going, to be persistent, to push to the end.

Many generous folks helped keep me going while writing this book. I wish to thank first the incredible Mark Frazier Lloyd, archivist emeritus at the University of Pennsylvania, who provided an astonishing amount of information on the Philadelphia medical scene in America's early years. He immediately and repeatedly shared his fount of knowledge on William Shippen and William Morgan, and also read and commented on a large section of content, which made two chapters considerably stronger. Even better, he was a joy to talk with.

Thanks as well to three archivists who proved similarly helpful as I plodded my way through the early chapters: Kathryn Hurwitz, archivist at Trinity Church in New York City; Stephen E. Novak, head of Archives & Special Collections at the Augustus C. Long Health Sciences Library at Columbia University; and Jocelyn K. Wilk, archivist at Columbia University's Rare Book & Manuscript Library, all of whom provided detailed answers to many of my probably inane questions. I wish also to thank Tom Hennessey, curator of the Lock Museum of America in Terryville, Pennsylvania; Douglas Arbittier, curator of the Arbittier Museum of Medical History in Mendham, New Jersey; and Jerry Kemp, owner of Cures AllDiseases.com.

I am grateful for clarifications and research suggestions from Dawnie Wolf Steadman, director of the Forensic Anthropology Center at the University of Tennessee; Michael Sappol, author of the acclaimed *A Traffic of Dead Bodies: Anatomy and Embodied Social Identity in Nineteenth-Century America*; and Susan Standring, emeritus professor of anatomy at King's College London.

Given the chance to interview experts, I would take that chance every time. Thank you first to Dawn Lynch, a physical therapist who helped straighten out a leg issue I was having and told me stories about Ulysses, the donor whose body helped Dawn learn muscle anatomy. Thank you as well to Ann Zumwalt, Carrie Rowell, and Angela McArthur, who took time out of their schedules to chat with me about anatomy education, embalming, and body donations, respectively. I very much enjoyed talking with the delightful Per-Olaf Hasselgren, the George H. A. Clowes Distinguished Professor of Surgery at Beth Israel Deaconess Medical Center, about smallpox inoculations in the eighteenth century.

For later chapters, I am most indebted to Thomas Champney and Paul Micah Johnson. Dr. Champney, anatomy professor at the University of Miami and director of the South Florida Willed Body Program for the State Anatomical Board, proved exceedingly helpful in my attempt to better understand body donations and the lack of governmental oversight of the body donation market. Tom has been a leader in those regards, and I am humbled to have been able to pick his learned brain. Likewise Agent Johnson, who provided insightful comments on the law enforcement side of body brokers. As a long-time investigator in the Detroit office of the FBI, Paul had been investigating what he called the "vast gray and black market of dead human bodies" for many years. He graciously imparted his boots-on-the-ground knowledge about that market, and for that I am most appreciative.

I am, as always, most indebted to my darling wife, Gay, whose patience with me while I spent day after day locked in my little office, pounding the keys and delving in the history of our nation and its medical past, was nothing short of miraculous. She recently discovered golf, a favorite hobby, and I regret not being able to play with her during her initial learning period. I was at the time trying to, as she often implored, "Finish that damn book!" Now that the manuscript is finished . . . *Fore!*

APPENDIX

MARY'S GHOST: A PATHETIC BALLAD

For the poetry lovers among us, here is the complete version of the poem from the epigraph, Thomas Hood's eerie ballad to Mary, a victim of body snatchers.

'Twas in the middle of the night,
To sleep young William tried,
When Mary's ghost came stealing in,
And stood by his bedside.

O William dear! O William dear!
My rest eternal ceases;
Alas! my everlasting peace
Is broken into pieces.

I thought the last of all my cares
Would end with my last minute;
But though I went to my long home,
I didn't stay long in it.

The body-snatchers they have come,
And make a snatch at me;
It's very hard them kind of men
Won't let a body be!

You thought that I was buried deep,
Quite decent-like and chary,
But from her grave in Mary-bone,
They've come and boned your Mary.

The arm that used to take your arm
Is took to Dr. Vyse;
And both my legs are gone to walk
The hospital at Guy's.

I vowed that you should have my hand,
But fate gives us denial;
You'll find it there, at Dr. Bell's,
In spirits and a phial.

As for my feet, the little feet
You used to call so pretty,
There's one, I know, in Bedford Row,
The t'other's in the City.

I can't tell where my head is gone,
But Doctor Carpue can;
As for my trunk, it's all packed up
To go by Pickford's van.

I wish you'd go to Mr. P.
And save me such a ride;
I don't half like the outside place,
They've took for my inside.

The cock it crows—I must be gone!
My William, we must part!
But I'll be yours in death, altho'
Sir Astley has my heart.

Don't go to weep upon my grave,
And think that there I be;
They haven't left an atom there
Of my anatomie.

Notes

Frontispiece

Wilson, T. C., "The dissecting room," *Center for the History of Medicine, Harvard Countway Library*, https://collections.countway.harvard.edu. Accessed February 29, 2024.

Author's Note

"'First,' said Swan, 'the first full reports . . .'" Robert J. Swan, "Prelude and Aftermath of the Doctors' Riot of 1788: A Religious Interpretation of White and Black Reaction to Grave Robbing," *New York History,* October 2000, 451.

Chapter 1: Medicine's Moral Conundrum

"A caricature from 1782 . . ." Unknown, 1782. "William Hunter (1718–1783) in His Museum in Windmill Street on the Day of Resurrection, Surrounded by Skeletons and Bodies, Some of Whom Are Searching for Their Missing Parts. Engraving, 1782." Wellcome Collection.

". . . having been appointed physician extraordinary . . ." William Munk, "William Hunter," *The Roll of the Royal College of Physicians of London* (London: The College Pall Mall East, 1878), Vol. II, 1701–1800, 207–8.

"Hunter typically supplied each anatomy student . . ." Toby Gelfand, "The 'Paris Manner' of Dissection: Student Anatomical Dissection in Early Eighteenth Century Paris," *Bulletin of the History of Medicine*, March–April 1972, 101.

"A writer in the Quaker journal . . ." Author signed as "Phillip's Vacation Thoughts," *The British Friends: Vol. 14, No. 1–12* (Glasgow: William and Robert Smeal, 1856), Vol. 8, 204.

"The first known case of body snatching . . ." Charles Singer, *A Short History of Anatomy & Physiology from the Greeks to Harvey* (New York: Dover Publications, 1957), 121.

"Leonardo da Vinci paid body snatchers . . ." Joseph K. Perloff, "Human Dissection and the Science and Art of Leonardo da Vinci," *American Journal of Cardiology*, January 31, 2013, www.ajconline.org. Accessed October 24, 2024.

"The great anatomist Andreas Vesalius . . ." Amber R. Comer, "The Evolving Ethics of Anatomy: Dissecting an Unethical Past in Order to Prepare for a Future of Ethical Anatomical Practice," *The Anatomical Record*, 2022, 820.

"Ruth Richardson, in her seminal work . . ." Ruth Richardson, *Death, Dissection, and the Destitute* (London: Penguin Books, 1988), 68.

"Wrote one physician in 1824 . . ." Suzanne M. Shultz, *Body Snatching: The Robbing of Graves for the Education of Physicians in Early Nineteenth Century America* (Jefferson, NC: McFarland, 2005), 20.

"The group, known as the 'Spunker Club . . .'" Nuriya Saifulina, "Harvard's Habeas Corpus: Grave Robbing at Harvard Medical School," *The Crimson*, www.thecrimson.com. Accessed Feb. 3, 2025.

"Warren once told of a body snatching . . ." Edward Warren, *The Life of John Warren, MD: Surgeon-General During the War of the Revolution, First Professor of Anatomy and Surgery in Harvard College, President of the Massachusetts Medical Society, Etc.* (Boston: Noyes, Holmes and Company, 1874), 233.

"According to famed British surgeon and anatomist Sir Astley Cooper . . ." James Blake Bailey, *The Diary of a Resurrectionist: To Which Are Added an Account of the Resurrection Men in London and a Short History of the Passing of the Anatomy Act, 1811–1812* (London: Swan Sonnenschein, 1896), 15.

"At a time when most American physicians . . ." Emails from Stephen Novak, March 31 and April 1, 2022.

"'We could not venture to meddle . . .'" Michael Sappol, *A Traffic of Dead Bodies: Anatomy and Embodied Social Identity in Nineteenth-Century America* (Princeton, NJ: Princeton University Press, 2002), 103.

"Students go in, first day of anatomy . . ." Interview with Ann Zumwalt, PhD, March 30, 2022.

"Between April 1788 and the end of 1884 . . ." Sappol, *Traffic*, 106.

"Riots also broke out in . . ." Linden F. Edwards, "Resurrection Riots During the Heroic Age of Anatomy in America," *Bulletin of the History of Medicine*, Mar.-Apr. 1951, 180–82.

"Castleton Medical College in Vermont decreed . . ." Arthur M. Lassek, *Human Dissection: Its Drama and Struggle* (Springfield, IL: Charles C Thomas, 1958), 209–10.

"No arrests were made . . ." Mark Bushnell, "Then Again: In Early Vermont, Robbery Was a Grave Concern," *VT Digger.com*, https://vtdigger.org. Accessed Jan. 18, 2025.

"'Grave robbing, naturally enough . . .'" "Body Snatching in Indianapolis," *Medical Standard*, vol. 25, no. 12, December 1902, 644.

"The bones were then sold . . ." "Alistair Cooke's Bones Were Stolen for Implantation, His Family Says," *New York Times*, Dec. 23, 2005, www.nytimes.com.

"The case became the subject . . ." Toby Dye, director, *Bodysnatchers of New York,* Filmrise, 2010.

"An embalmer from Staten Island . . ." "Funeral Parlor Owner from Staten Island Sentenced in Body Parts Scheme," *Staten Island Advance,* March 21, 2019, www.silive .com/news/2009/01/former_staten_island_funeral_p.html.

"In the summer of 2023. . ." Laura Ly, "Former Harvard Medical School Morgue Manager Is Accused of Stealing, Selling and Shipping Human Body Parts, Indictment Says," *CNN,* www.cnn.com.

"'I remember walking by a dissection table . . .'" "Letters of Appreciation," *Division of Anatomy, Ohio State University College of Medicine,* https://medicine.osu.edu/. Accessed March 23, 2025.

"Our donors taught us . . ." Lena H. Sun, "Medical Students Learned on the Bodies, and Now Honor the Donors," *Washington Post,* May 27, 2012, www.washing tonpost.com.

CHAPTER 2: A NATION AT PEACE AND IN TURMOIL

"His great-grandfather, Jacobus Van Cortlandt . . ." "Jacobus van Cortlandt [1658–1739]: Early Founder/Historic Leader," *New Netherland Institute,* www .newnetherlandinstitute.org.

"'This work I mean to do at my own expense . . .'" "The American Commissioners of the Preliminary Peace Negotiations with Great Britain," *Diplomatic Reception Rooms, US Department of State,* www.diplomaticrooms.state.gov. Accessed April 6, 2025.

"The final painting was to include . . ." Poor health prevented Franklin from sitting for West, but a friend of Franklin's provided West with a painting of him as a reference.

"He was too ill on his release . . ." Walter Stahr, *John Jay: Founding Father* (New York: Diversion Books, 2012), 194.

"Oswald, however, had been forced to resign . . ." Ellen Tucker, "American Independence Begins on Favorable Terms: Continental Congress Ratifies the Treaty of Paris," *Teaching American History,* January 14, 2021, https://teachingamericanhistory .org. Accessed February 2, 2022.

"As I very strongly expressed my regret . . ." Artwork details *Metropolitan Museum of Art,* www.metmuseum.org. Accessed January 21, 2022.

"'American Raphael'" Shovava, "How American Benjamin West Became London's Preeminent Painter During the American Revolution," *My Modern Met,* https://my modernmet.com. Accessed April 14, 2022.

"Oscar Zeichner, professor emeritus . . ." Oscar Zeichner, "The Loyalist Problem in New York," *New York History,* July 1940, 287.

"'In fact, some of the big battles . . .'" "What Happened to British Loyalists After the Revolutionary War?" *All Things Considered,* NPR, July 3, 2015, www.npr.org. Accessed April 11, 2022.

"By 1783 some sixty- to eighty-thousand Loyalists . . ." Shannon Duffy, "Loyalists," *George Washington's Mount Vernon,* www.mountvernon.org. Accessed April 23, 2022.

"When the war began the colonies . . ." Edward Ayres, "African Americans and the American Revolution," Jamestown-Yorktown Foundation Museums, www .jyfmuseums.org. Accessed April 20, 2022.

"Virginia governor Lord Dunmore issued . . ." "Lord Dunmore's Proclamation (Transcription), *Museum of the American Revolution,* www.amrevmuseum.org. Accessed April 21, 2022.

"'New York City, occupied . . .'" David Kobrin, *The Black Minority in Early New York* (Albany, NY: New York State Education Department, 1971), 40.

"British historian Simon Schama explained . . ." Simon Schama, "Dirty Little Secret," *Smithsonian Magazine,* May 2006, www.smithsonianmag.com.

"English-speaking people called the alliance . . ." "Today," *Onondaga Nation,* www.onondaganation.org/aboutus/today. Accessed April 16, 2022.

"'While time and fortune has helped . . .'" William Sawyer, "The Six Nations Confederacy During the American Revolution," *National Park Service,* www.nps .gov. Accessed April 16, 2022.

"For instance the Forfeiture Act . . ." Joseph Tiedemann, "Loyalists and Conflict Resolution in Post-Revolutionary New York: Queens County as a Test Case," *New York History,* January 1987, 32.

"'Cursed, cursed Tyrants . . .'" Tiedemann, "Loyalists," 27.

"He railed against more extreme views . . ." Oscar Zeichner, "The Loyalist Problem in New York After the Revolution," *New York History,* July 1940, 288.

"'I used to tell him . . .'" Ron Chernow, *Alexander Hamilton* (New York: Penguin Publishing Group, 2004), 190.

"'He thought America's character . . .'" Chernow, *Hamilton,* 195.

"So many prisoners died each day . . ." Charles H. Jenrich, "The Old Jersey Prison Ship," US Naval Institute Proceedings, February 1963, www.usni.org. Accessed June 17, 2024.

"One writer claimed that George Clinton . . ." Oscar Zeichner, "The Loyalist Problem in New York After the Revolution," *New York History,* July 1940, 286.

"Writing under the pen name Phocion . . ." "A Letter from Phocion to the Considerate Citizens of New York, [1–27 January 1784]," *Founders Online, National Archives,* https://founders.archives.gov. Accessed October 26, 2023. (Original source: *The Papers of Alexander Hamilton,* Vol. 3, 1782–1786, ed. Harold C. Syrett [New York: Columbia University Press, 1962], 483–97.)

"Hamilton further asserted . . ." "Draft Peace Treaty Presented by Richard Oswald to the American Peace Commissioners, 25 November 1782," *Founders Online,* https://founders.archives.gov/documents/Adams/06-14-02-0048.

"'The fact is,' wrote Hamilton . . ." Alexander Hamilton, "From Alexander Hamilton to George Washington, 9 July 1795," *Hamilton Papers,* https://founders .archives.gov. Accessed April 6, 2025.

"In perhaps the most famous of the Trespass Act cases . . ." Henry B. Dawson, ed., *The Case of Elizabeth Rutgers Versus Joshua Waddington, Determined in the Mayor's Court, in the City of New York, August 7, 1786* (New York: Morrisania, 1866), xix.

"When the British took over . . ." Chernow, *Hamilton,* 197.

"Her team also included . . ." Benjamin F. Thompson, *History of Long Island* (New York: E. French, 1839), 408.

"Livingston was a classmate of James Madison . . ." "(Henry) Brockholst Livingston," *Historical Society of the New York Courts,* https://history.nycourts.gov. Accessed October 20, 2022.

"Lewis began working in Jay's law office . . ." "Morgan Lewis," *Historical Society of the New York Courts,* https://history.nycourts.gov. Accessed April 6, 2025.

"John Adams once described Duane . . ." Ron Chernow, *Alexander Hamilton* (New York: Penguin Publishing Group, 2004), 199.

"Spectators in the courtroom . . ." Chernow, *Hamilton,* 198.

"'Duane suggested . . .'" Peter Charles Hoffer, *Rutgers v. Waddington: Alexander Hamilton, the End of the War for Independence, and the Origins of Judicial Review* (Lawrence, KS: University Press of Kansas, 2016), 79.

CHAPTER 3: LIFE IN THE NEW REPUBLIC

"Kariann Akemi Yokota, in her illuminating book . . ." Kariann Akemi Yokota, *Unbecoming British: How Revolutionary America Became a Post-Colonial Nation* (New York: Oxford University Press, 2011), 12.

"'Their words were sometimes those of age to youth . . .'" C. Vann Woodward, *The Old World's New World* (Oxford: Oxford University Press, 1992), vii.

"'Colonial Americans,' wrote Joseph J. Ellis . . ." Joseph J. Ellis, *After the Revolution: Profiles of Early American Culture* (New York: W. W. Norton, 2002), 12. Kindle Edition.

"Ben Franklin once expressed his jealousy . . ." E. N. da C. Andrade, "Benjamin Franklin in London," *Journal of the Royal Society of Arts,* February 3, 1956, 231.

"King's College was established . . ." Donald O. Quest, "The Origins of the College of Physicians and Surgeons of Columbia University," *History of Medicine,* Summer 1997, 137.

"The first classes, held in a school house . . ." "The History of Columbia College," *Columbia College,* www.college.columbia.edu. Accessed July 24, 2022.

"'The chief thing that is aimed at in this college . . .'" Robert McCaughey, excerpt from *Stand, Columbia: The Founding of King's College* (New York: Columbia University Press, 2003), www.college.columbia.edu. Accessed April 7, 2025.

"The college moved about two years . . ." John B. Pine, "King's College and the Early Days of Columbia College," *Proceedings of the New York State Historical Association*, 17 (1919), 116.

"The college moved to a 'stately . . .'" Chernow, *Hamilton*, 49.

"The building and its grounds . . ." Chernow, *Hamilton*, 49.

". . . send a fleet of warships . . ." Chernow, *Hamilton*, 119. Lafayette's navy arrived in July the year Hamilton graduated.

". . . had received his inclination toward medicine . . ." Samuel Bard and his father, John, would one day remove a "very large and painful tumor" on a newly inaugurated President George Washington's thigh. The tumor would today be called a carbuncle. James E. Guba, "Anthrax and the President, 1789," *Washington Papers*, Spring 2002, https://washingtonpapers.org.

"I feel a little jealous of the Philadelphians . . ." "It Happened Here: Dr. Samuel Bard," *New York–Presbyterian Health Matters*, https://healthmatters.nyp.org. Accessed January 16, 2022.

"When Samuel returned home to New York . . ." "Samuel Bard and the King's College School," *New York Academy of Medicine*, May 1925, 87.

". . . including Peter Middleton, a Scottish physician . . ." Thomas Gallagher, *A Doctor's Story: In Commemoration of the 200th Anniversary of the Columbia University College of Physicians and Surgeons* (New York: Harcourt, Brace & World, 1967), 381.

". . . including Peter Middleton . . ." "The Exhibits of the Smithsonian Institution and United States National Museum at the Alaska-Yukon-Pacific Exposition, Seattle, Washington, 1909," *Smithsonian Institution* (Boston: Press of Judd & Detweiler, 1909), 98.

". . . who had published a seminal work . . ." Byron Stookey, "Samuel Clossy, AB, MD, FRCP of Ireland: First Professor of Anatomy, King's College (Columbia), New York," *Bulletin of the History of Medicine*, 38, 1964, 153.

". . . and John Jones, a pioneering surgeon . . ." Mark A. Hardy, "John Jones, MD: Pioneer, Patriot, and Founder of American Surgery," *World Journal of Surgery*, April 2010, 605.

"Bard's efforts proved successful . . ." "New York-Presbyterian Hospital/Weill Cornell," *Weill Cornell Medicine Samuel J. Wood Library*, https://library.weill.cornell.edu. Accessed July 25, 2022.

"Hospitals were breeding grounds . . ." Lindsay Fitzharris, "Appointment at the House of Death: The Horror of the Early Victorian Hospital," *History Extra*, www.historyextra.com. Accessed October 20, 2022.

"Jones had visited hospital wards in Paris . . ." Francis Randolph Packard, *The History of Medicine in the United States* (Philadelphia: J. B. Lippincott, 1901), 362–63.

"'The principal wards . . .'" Packard, *Medicine*, 262.

"A cornerstone for New York Hospital . . ." Quest, *Origins*, 140.

". . . set about thirty yards back . . ." Wood Library, *New York-Presbyterian*.

". . . when the building took fire . . ." Email from Chiyong (Tali) Han, archivist at Archivist at the Medical Center Archives for New York-Presbyterian/Weill Cornell Medicine, quoting Henry Crane's *History of the Society of the New York Hospital Compiled from its Records, 1769 to 1872* (New York: New York Hospital Society, 1872), 38. Received October 21, 2022.

"A surgeon's mate named Samuel Drowne . . ." Marynita Anderson, *Physician Heal Thyself: Medical Practitioners of Eighteenth Century New York* (New York: Peter Lang Group, 2004), 90.

"A surgeon from Rhode Island . . ." Byron Stookey, *A History of Colonial Medical Education in the Province of New York, With Its Subsequent Development (1767–1830)* (Springfield, IL: Charles C Thomas Publisher, 1962), 74–75.

"Drowne nearly needed his own services . . ." Eric Larrabee, *The Benevolent and Necessary Institution: The New York Hospital, 1771–1971* (Garden City, NY: Doubleday & Co., 1971), 87.

"A month later the British Army seized . . ." Wood Library, *New York-Presbyterian*.

"James Duane, New York's Whig mayor . . ." *Columbia University Quarterly*, March 1909, 155.

"John B. Pine, a long-time trustee . . ." John B. Pine, "King's College and the Early Days of Columbia College," *Proceedings of the New York State Historical Association*, 1919, 121.

Columbia College is today the undergraduate school of Columbia University.

"The ship landed at Staten Island . . ." "Dr. Richard Bayley: Physician, Educator, and Researcher," *Seton Hall University*, https://blogs.shu.edu. Accessed January 23, 2024.

"The only way he could visit her . . ." Davina Waterson, "Richard Bayley (1745–1801)," *Eclectic Medical Journal*, June 1917, 312.

"So incensed were the accusers . . ." *William S. Smith to George Clinton, 20 October 1783. Public papers of George Clinton, First Governor of New York, 1777–1795* (Albany, NY: The State of New York, 1904), Vol. 8, 265.

"'When I first saw Doctor Bayley . . .'" *Smith to Clinton*, 265.

"His name now cleared Bayley . . ." Michael Vigorito, "Dr. Richard Bayley: Physician, Educator, and Researcher," excerpt from presentation to Staten Island in American History and 21st-Century Education, Education Symposium/Academic Conference, College of Staten Island-CUNY, March 19–20, 2011.

"His brother, John, who became . . ." "Dr. William Hunter," *The Huntarian*, www.gla.ac.uk. Accessed February 12, 2024.

"'Men of the world, artists, . . .'" William Macmichael, *Lives of British Physicians* (London: John Murray, 1839), 228.

"Innumerable colleagues would have agreed . . ." William Munk, "William Hunter," *Royal College of Physicians*, http://history.rcplondon.ac.uk/. Accessed January 19, 2024. Anyone who writes about medical history in the United Kingdom owes a debt to William Munk for compiling what is commonly known as the Munk Roll, a compendium of short biographies of UK physicians. Published first in 1855 and initially consisting of three volumes, the Roll provided detailed information on physicians in practice between 1518 and 1800. The Roll today continues to be maintained and updated by the Royal College of Physicians.

"His collection had grown so large . . ." *The Hunterian*.

"In it was 'a handsome amphitheater . . .'" Macmichael, *British Physicians*, 226.

"Cartoonists joined in . . ." *The Hunterian*.

"Du Verney studied diseases and disorders of the ear . . ." H. Dominic W. Stiles, "Guichard Duvernay, Pioneer of Otology (1648–1730)," *University College London*, blogs.ucl.ac.uk. Accessed April 7, 2025.

"Guichard Joseph Du Verney, a pioneering . . ." Regis Olry, "Body Snatchers: The Hidden Side of the History of Anatomy," *Journal of the International Society for Plasti-nation*, vol. 14, no. 2 (1999): 7.

"A relative of Sterne's . . ." Olry, "Body Snatchers," 7.

"I have often heard [Hamilton] speak . . ." Chernow, *Hamilton*, 52.

"These were stirring days for Hamilton . . ." Chernow, *Hamilton*, 55.

"He supported a boycott of British goods . . ." Broadus Mitchell, *Alexander Hamilton: Youth to Maturity* (New York: Macmillan Co., 1957), 64.

"He proclaimed that without that unity . . ." Mitchell, *Youth*, 63.

"Born into a military family . . ." Reneé Critcher Lyons, *Foreign-Born American Patriots: Sixteen Volunteer Leaders in the Revolutionary War* (Jefferson, NC: McFarland, 2014), 190.

"He gave the members a letter . . ." "Baron Von Steuben," *National Park Service, Valley Forge National Park*, www.nps.gov. Accessed March 4, 2024.

"He lost his commission . . ." Erick Trickey, "The Prussian Nobleman Who Helped Save the American Revolution," *Smithsonian Magazine*, April 26, 2017, www.smithsonianmag.com. Accessed January 26, 2022.

"No evidence to support that rumor . . ." Paul Douglas Lockhart, *The Drillmaster of Valley Forge: The Baron de Steuben and the Making of the American Army* (New York: HarperCollins, 2007), 42.

"Rather than stay and provide a defense . . .'" Erin Blakemore, "The Revolutionary War Hero Who Was Openly Gay," *History.com*, www.history.com. Accessed January 26, 2022.

"With that support . . ." "Baron von Steuben," *Battlefields.org*, www.battlefields .org. Accessed January 26, 2022.

"The trappings of his horse . . .'" General von Steuben," *Valley Forge National Historical Park*, www.nps.gov. Accessed January 26, 2022.

"Washington was also impressed . . ." Blakemore, *Gay*.

"Steuben spoke French but little English . . ." Blakemore, *Gay*.

"'My dear Duponceau . . .'" *Battlefields*.

"Steuben greatly improved sanitation . . ." "Biography of Baron von Steuben," *USHistory.com*. Accessed January 26, 2022.

"He resigned his commission . . ." Joseph B. Doyle, *Frederick William Von Steuben and the American Revolution* (Steubenville, OH: H. C. Cook Company, 1913), 328.

"Several friends visited . . ." Blakemore, "Openly Gay."

"Biographer Joseph Doyle said that 'Steuben . . .'" Doyle, *Steuben*, 330.

". . . took his meals with Misses Dabeny . . ." Doyle, *Steuben*, 330.

CHAPTER 4: DISSECTION THROUGH THE AGES

"'And this old man . . .'" "The Veins of the Arm; And Notes on the Death of a Centenarian c. 1508," *Royal Trust Collection*, www.rct.uk/collection. Accessed June 18, 2024.

"When he died his body . . ." Roger Jones, "Leonardo da Vinci: Anatomist," *British Journal of General Practice*, June 2012, 319.

". . . 'desiccated and like congealed bran . . .'" Marton Clayton, "Medicine: Leonardo's Anatomy Years," *Nature*, April 18, 2012, 315.

". . . 'falls away in tiny flakes . . .'" Royal Trust Collection, *Centenarian*.

"The results are triumphs . . ." Walter Isaacson, *Leonardo da Vinci* (New York: Simon & Schuster, 2017), 523.

"Post-mortem dissection was allowed . . ." Krishan M. Thadani, "The Myth of a Catholic Religious Objection to Autopsy," *National Catholic Bioethics Center*, 2012, 41.

"The subjects came from prisons . . ." "Anatomical Theaters," *Museum of the History of Medicine at the Medical University of Warsaw*, https://teatranatomiczny.wum.edu.pl. Accessed April 7, 2024.

"'Anatomy typically had the blessings . . .'" Michael Sappol, "Sappol Responds," *American Journal of Public Health*, March 2003, 364.

"'He and his students forged keys . . .'" Katherine Park, "The Criminal and the Saintly Body: Autopsy and Dissection in Renaissance Italy," *Renaissance Quarterly*, Spring 1994, 18.

"Park wrote that 'he opened the body . . .'" Park, "Criminal and Saintly," 19.

"He hated the normal manner of conducting . . ." Fabio Zampieri, Mohamed ElMaghawry et al., "Andreas Vesalius: Celebrating 500 years of Dissecting Nature," *Global Cardiology Science & Practice*, December 22, 2015, www.qscience.com. Accessed April 9, 2025.

"Vesalius decried this 'hateful method . . .'" Zampieril, "Celebrating."

"An assistant to the pope . . ." Alisha Rankin, "Most Execrable and Abominable or Irreligious," *Lapham's Quarterly*, January 14, 2021, www.laphamsquarterly.org. Accessed April 8, 2024.

"Her fellow nuns decided . . ." Park, *Saintly Bodies*, 1.

"Women were considered particularly prone . . ." Kalina Yamboliev, 2010, "Blurring the Lines: Private and Public Dissection in Renaissance Italy" (Unpublished bachelor's thesis, 2010), University of Nevada, Reno, 5.

"Viewers examining her heart found three small stones . . ." Yamboliev, "Blurring the Lines," 6.

"Pope Clement VII officially decreed . . ." Sanjib Kumar Ghosh, "Human Cadaveric Dissection: A Historical Account from Ancient Greece to the Modern Era," *Anatomy and Cell Biology*, 2015, 158, https://pmc.ncbi.nlm.nih.gov. Accessed April 10, 2024.

"Each should be a sizeable and well-ventilated . . ." Gert-Horst Schumacher, "*Theatrum Anatomicum* in History and Today," *International Journal of Morphology*, vol. 25, 2007, 21, www.scielo.cl. Accessed April 15, 2024.

"The Leiden theater, for instance, was . . ." Tim Huisman, "The Finger of God: Anatomical Practice in 17th-Century Leiden," PhD Diss. (University of Leiden, 2008), 10, https://hdl.handle.net. Accessed April 7, 2024.

"The latter part of the fifteenth century . . ." Schumacher, "*Theatrum Anatomicum*," 21.

"Members of the paying-public . . ." Howard Hotson, "Anatomical Theaters: Padua, 1594," *Cabinet*, www.cabinet.ox.ac.uk/anatomical-theatres-padua-1594. Accessed April 16, 2024.

"The permission came, however . . ." Arthur M. Lassek, *Human Dissection: Its Drama and Struggle* (Springfield, IL: Charles C Thomas, 1958), 155.

"Many of them wanted to study . . ." Robert Chambers and Thomas Napier Thomson, Ed., *A Biographical Dictionary of Eminent Scotsmen, Vol. 6* (Glasgow: Blackie and Son, 1857), 37–44.

"During this time physicians and surgeons . . ." Lassek, *Human Dissection*, 116.

"Percivall Pott, the first professor . . ." Carolina Abboud, "Percival Pott (1714–1788)," *Arizona State University Embryo Project Encyclopedia*, https://embryo.asu.edu. Accessed April 22, 2024.

"During the 1700s in England . . ." Ghosh, "Human Cadaveric Dissection," 158.

"The body of criminals convicted and executed . . ." "1751: 25 George 2 c., 37: The Murder Act," *The Statutes Project: Putting Historic British Law Online*, https://statutes.org.uk. Accessed April 3, 2024.

"So many bodies of executed criminals . . ." Ghosh, "Cadaveric Dissection," 160.

"The act further stipulated . . ." Statutes.org, *Murder Act*.

"Historians differ about the first dissection in the colonies, with one . . ." Lassek, *Human Dissection*, 184.

"Packard said that a man named ..." Francis Randolph Packard, *History of Medicine in the United States* (Philadelphia: J. B. Lippincott, 1901), 62.

"However, from early legal codes ..." Lassek, *Human Dissection*, 182–83.

"The first European medical man ..." "Tooth Puller," *Historic Jamestown*, https://historicjamestowne.org. Accessed April 28, 2024.

"He wanted to be accepted ..." Kerry Falvey, "The Colonies' First Medical Degree," *Yale Medicine Magazine*, Winter 2010, https://medicine.yale.edu. Accessed April 24, 2024.

"'If Your Lordships judge me worthy ...'" Falvey, "First Degree."

"Recognizing their divergent skills and goals ..." "History of the Company," *Barber's Company*, https://barberscompany.org. Accessed August 19, 2024.

"'In the England of 1700 ...'" Richard H. Shryock, "Eighteenth Century Medicine in America," *Proceedings of the American Antiquarian Society*, October 1948, 279, www.americanantiquarian.org. Accessed December 24, 2024.

"Another medical historian, Irvine Loudon ..." Irvine Loudon, "Why Are (Male) Surgeons Still Addressed as Mr?" *British Medical Journal*, December 23–30, 2000, 1589, DOI: 10.1136/bmj.321.7276.1589. Accessed August 20, 2024.

CHAPTER 5: MEDICAL MELEE

"Shippen biographer Betsy Copping Corner ..." Betsy Copping Corner, *William Shippen, Jr.: Pioneer in American Medical Education* (Philadelphia: American Philosophical Society, 1951), 3.

"During his time in London Shippen decided ..." Corner, *Shippen*, 52.

"Shippen decided in early 1760 ..." Corner, *Shippen*, 82.

"He left an inscription ..." Whitfield J. Bell Jr., *John Morgan: Continental Doctor* (Philadelphia: University of Pennsylvania Press, 1965), 19.

"He brought with him ..." Bell, *Morgan*, 44.

"There he studied with William Hunter ..." Betsy Copping Corner, "Dr. John Fothergill and the American Colonies," *Quaker History*, Autumn 1963, 79, www.jstor.org/stable/41946415. Accessed July 14, 2024.

"'Dr. Shippen (who arriv'd from Scotland ...'" Bell, *Morgan*, 52.

"'Morgan inquired eagerly about Edinburgh ...'" Bell, *Morgan*, 52.

"That was all Morgan needed to hear ..." Bell, *Morgan*, 52.

"Morgan traveled to Rotterdam ..." Bell, *Morgan*, 91.

"He became the first American fellow ..." Bell, *Morgan*, 105.

"Students would learn human anatomy ..." Tiffany DeRewal, 2020, "The Resurrection and the Knife: Protestantism, Nationalism, and the Contest for the Corpse During the Rise of American Medicine," PhD Diss., Temple University, 44.

"For the city being small, almost everyone ..." William E. Horner, *Introductory Lecture to a Course of Anatomy in the University of Pennsylvania* (Philadelphia: J. G. Auner, 1831), 10.

"A member of one of the political parties . . ." William T. Parson, "The Bloody Election of 1742," *Pennsylvania History: A Journal of Mid-Atlantic Studies*, July 1969, 303.

"The lab at which he conducted his anatomy classes . . ." Phone interview with Mark Frazier Lloyd, June 11, 2024.

"Aware of the rumors . . ." No specific date of the Sailor's Riot, sometimes called the Sailor's Mob, has been identified. It seems most probable that it occurred after one of Shippen's ads appeared listing him as "Professor of Anatomy" at the new medical school.

"'It has given Dr. Shippen much pain . . .'" Advertisement in *The Pennsylvania Gazette*, September 26, 1765, page 1, https://www.newspapers.com/image/39405577. Accessed July 13, 2024.

"The trustees approved the plan at the same meeting, on May 3 . . ." Joseph Carson, *A History of the Medical Department of the University of Pennsylvania* (Philadelphia: Lindsay and Blakiston: 1869), 53.

"The trustees approved the plan and . . ." *200 Years of American Medicine* (Bethesda, MD, exhibit of *National Library of Medicine Publications and Productions*, 1976), page numbers not available.

"He then quickly added . . ." Bell, *Morgan*, 33–34.

"'Smith was a man of self-aggrandizement . . .'" Lloyd, Interview.

"'The managers had no interest in . . .'" Stephen Fried, *Rush: Revolution, Madness, and Benjamin Rush, the Visionary Doctor Who Became a Founding Father* (New York: Crown Publishing Group, 2019), 42.

". . . graduating in 1757 at age fifteen . . ." Fried, *Rush*, 40.

"Prior to his return to America Shippen . . ." William Shainline Middleton, "John Morgan, Father of Medical Education in North America," *Annals of Medical History*, March 1927, 16, https://pmc.ncbi.nlm.nih.gov. Accessed August 13, 2024.

"Historians have variously described Shippen . . ." Arthur M. Lassek, *Human Dissection: Its Drama and Struggle* (Springfield, IL: Charles C Thomas, 1958), 217.

". . . as well as 'bright, overbearing . . .'" Fried, *Rush*, 37.

"Shippen has been described . . ." Lassek, *Human Dissection*, 217.

"Two weeks later Shippen wrote . . ." Carson, *Medical Department*, 55–56.

"Ever tactful, Shippen ended with . . ." Carson, *Medical Department*, 55–56.

"Shippen soon became . . ." Fried, *Rush*, 40.

"As a keen reader of newspapers . . .'" Fried, *Rush*, 43.

"Morgan had told Rush before . . ." Fried, *Rush*, 49.

"'Dr. Shippen was my oldest friend . . .'" *Letters of Benjamin Rush: Volume I: 1761–1792* (Princeton, NJ: Princeton University Press, 2019), 62.

"Said Fried, 'Since Shippen left almost no writing . . .'" Fried, *Rush*, 83.

". . . Benjamin Church, who had been dismissed . . ." "Benjamin Church," *Intelligence.gov*, www.intelligence.gov. Accessed May 7, 2024.

"'Like Morgan before him . . .'" Lassek, *Human Dissection*, 220.

"Rush held no fondness for Shippen . . ." L. H. Butterfield, Ed., *Letters of Benjamin Rush: Volume I: 1761–1792* (Princeton, NJ: *Princeton University Press*, 1951), 161.

CHAPTER 6: BULLETS, BLOODLETTING, AND BAYONETS

"Clergy believed that . . ." Per-Olof Hasselgren, "The Smallpox Epidemics in America in the 1700s and the Role of the Surgeons: Lessons to Be Learned During the Global Outbreak of COVID-19," *World Journal of Surgery*, 2020, 2838.

"Boston, for instance, hadn't seen a case . . ." Francis Randolph Packard, *The History of Medicine in the United States* (Philadelphia: J. B. Lippincott, 1901), 77.

"Francis Packard, noted Philadelphia physician . . ." Packard, *Medicine in the US*, 77.

"One physician who had been trained . . ." Margot Minardi, "Boston Inoculation Controversy of 1721," *William and Mary Quarterly*, January 2004, https://www.jstor.org/stable/3491675, 49. Accessed July 24, 2024.

"'When, however, he communicated . . .'" Packard, *Medicine in US*, 77.

"'Indeed, so strong was the opposition . . .'" Hasselgren, "Smallpox," 2839.

"'Many pious, respectable people . . .'" Packard, *Medicine in the US*, 78.

"Someone tossed a small bomb . . ." Mark Best et al., "'Cotton Mather, You Dog, Dam You! I'l Inoculate You with This; With a Pox to You': Smallpox Inoculation, Boston, 1721," *Quality & Safety in Health Care*, 2004, DOI: 10.1136/qshc.2003.008797, 82. Accessed July 19, 2024.

"It read, 'Cotton Mather . . .'" Hasselgren, "Smallpox," 2839.

"Boylston ended up inoculating . . ." Minardi, *Boston*, 57.

"Boylston ended up inoculating . . ." Arthur Boylston and A. E. Williams, "Zabdiel Boylston's Evaluation of Inoculation Against Smallpox," *Journal of the Royal Society of Medicine*, September 1, 2008, DOI: 10.1258/jrsm.2008.08k008, 476. Accessed July 19, 2024.

"Boylston was not formally trained . . ." Packard, *Medicine in the US*, 81.

"Benjamin was sent to learn . . ." Horace Davis, *Dr. Benjamin Gott: A Family of Doctors* (Cambridge: Cambridge University Press, 1909), reprinted from the publications of the Colonial Society of Massachusetts, Vol. 12, 214.

"Baker's parents promised that . . ." Davis, *Gott*, 216.

"Baker, like most apprentices . . ." Abraham Flexner, *Medical Education in the United States and Canada: A Report to the Carnegie Foundation For the Advancement of Teaching* (Boston: Merrymount Press, 1910), 3.

"Only about ten percent of an estimated . . ." Liliana Camison et al., "The History of Surgical Education in the United States: Past, Present, and Future," *Annals of Surgery*, March 2022, DOI: 10.1097/AS9.0000000000000148, 1. Accessed August 16, 2024.

"Politician and historian William Smith . . ." Byron Stookey, *A History of Colonial Medical Education: In the Province of New York, with Its Subsequent Development (1767–1830)* (Springfield, IL: Charles C Thomas Publisher, 1962), 3.

"There is no city in the world . . ." Stookey, *Medical Education,* 4.

"Virginia had taken a swipe . . ." Wyndham B. Blanton, *Medicine in Virginia in the Seventeenth Century* (New York: Arno Press and the New York Times, 1972), 256.

"Said one prominent Harvard-trained pathologist . . ." Reginald H. Fitz, "Annual Oration 1894," Address to the Massachusetts Medical Society, www.massmed.org. Accessed July 24, 2024.

"One Philadelphia wigmaker . . ." Nissa M. Strottman, "Public Health and Private Medicine: Regulation in Colonial and Early National America," *Hastings Law Journal,* vol. 50, no. 2, 1999, https://repository.uchastings.edu/hastings_law _journal/vol50/iss2/4, 400. Accessed April 7, 2025.

"Nissa Strottman, an attorney who studied regulations . . ." Strottman, "Public Health," 401.

"'The Widow Read,' said one advertisement . . .'" "Extracts from the *Gazette,* 1731," *Founders Online,* https://founders.archives.gov/documents/Franklin/01-01 -02-0069. Accessed July 28, 2024.

"Numerous patent medicines . . ." George B. Griffenhagen and James Harvey Young, "Old English Patent Medicines in America," *Pharmacy in History,* vol. 34, no. 4, 1992, 200.

"While in London in 1737 he persuaded three physicians . . ." Wyndham B. Blanton, *Medicine in Virginia in the Eighteenth Century* (Richmond, VA: Garrett & Massie Inc., 1980), 122.

"Historian Richard M. Jellison . . ." Richard M. Jellison, "Dr. John Tennant and the Universal Specific," *Bulletin of the History of Medicine,* July–August 1963, 336.

"One calling himself 'I. G.' . . ." Jellison, "Tennant," 340.

"Tennant's belief in Seneca snakeroot . . .'" Jellison, "Tennant," 346.

"Mather added a perhaps unnecessary codicil . . ." Richard H. Shryock, "Eighteenth Century Medicine in America," *Proceedings of the American Antiquarian Society,* October 1948, 275.

"One substance finding widespread favor . . ." Mary C. Gillett, *The Army Medical Department, 1775–1818* (Washington, DC: Center of Military History, United States Army, 2004), 7.

"One noted physician of the time . . ." Gillett, *Army,* 7.

"Bloodletting, in the words of one Medieval writer . . ." Lynn Thorndike, *A History of Magic and Experimental Science During the First Thirteen Centuries of Our Era, Vol. 1* (New York: Columbia University Press, 1923), 728.

"Spring-loaded fleams made more . . ." Timothy M. Bell, "A Brief History of Bloodletting," *Journal of Lancaster General Hospital,* Winter 2016, 119.

"Because leeches leave behind a mild anticoagulant . . ." Ralph G. DePalma et al., "Bloodletting: Past and Present," *Journal of the American College of Surgeons*, 136.

"Following a seizure and . . ." Jeffrey K. Aronson and Carl Heneghan, "The Death of King Charles II," *University of Oxford Centre for Evidence-Based Medicine*, www .cebm.ox.ac.uk. Accessed June 27, 2024.

"'When the doctor was called in . . .'" Simon Jong-Koo Lee, "Infective Endocarditis and Phlebotomies May Have Killed Mozart," *Korean Circulation Journal*, 2010, 611.

"Washington received an enema . . ." Ron Chernow, *Washington: A Life* (New York: Penguin Publishing Group, 2011), 807.

"At dinner time on December 14 . . ." Tobias Lear, "I, 15 December 1799," *Founders Online*, https://founders.archives.gov/documents/Washington/06-04 -02-0406-0001.

"Washington died a few hours later." Physicians and historians today differ on the cause of his death, putting forth such maladies as diphtheria, Ludwig's angina (dangerous bacterial infection of the mouth and throat), pneumonia, quinsy (tonsillar abscess), strep throat, Vincent's angina (bacterial infection of the mouth), and, most recently, acute epiglottitis as the culprit. We may never know the true cause.

"In it, he recommended bleeding for most gunshot wounds . . ." Mark M. Ravitch, "Surgery in 1776," *Annals of Surgery*, September 1977, 293.

"He also came to believe that there existed . . ." Mary C. Gillett, *The Army Medical Department, 1775–1818* (Washington, DC: Center of Military History, US Army, 2004), 2.

"Cobbett had reviewed mortality records . . ." Robert L. North, "Benjamin Rush, MD: Assassin or Beloved Healer?" *Baylor University Medical Center Proceedings*, vol. 13, no. 1, 48.

"The Israelite slew his thousands . . ." North, "Assassin," 48.

"He concluded that his results . . ." Alfredo Morabia, "Pierre-Charles-Alexandre Louis and the Evaluation of Bloodletting," *Journal of the Royal Society of Medicine*, March 2006, 159.

"Gradually, as other studies examined bloodletting . . ." A form of bloodletting, or phlebotomy, occurs even today, though by far safer means and only for an extremely limited number of conditions, including sickle cell anemia; polycythemia vera, a disorder in which the body produces an abnormally high number of red blood cells; and hemochromatosis, in which the body absorbs too much iron from foods.

"Renowned Harvard surgeon . . ." Henry J. Bigelow, "A History of the Discovery of Modern Anesthesia," in *A Century of American Medicine, 1776–1876*, Edward H. Clark et al. (Philadelphia: Henry C. Lea, 1876), 79.

"A thirteenth-century monk named Theodoric . . ." Mohamad Said Maani Takrouri, "Historical Essay: An Arabic Surgeon, Ibn al Quff's (1232–1286) Account on Surgical Pain Relief," *Anesthesia: Essays and Researches*, January–June 2010, www.ncbi .nlm.nih.gov/pmc/articles/PMC4173333. Accessed August 14, 2024.

"Ether, invented in 1540 . . ." Connie Y. Chang et al., "Ether in the Developing World: Rethinking an Abandoned Agent," *BMC Anesthesiology*, October 16, 2015, https://bmcanesthesiol.biomedcentral.com. Accessed August 14, 2024.

"'American students adopted a variation . . .'" Chang, *Ether*.

"'The conditions under which the surgeons toiled . . .'" Richard L. Blanco, "Military Medicine in Northern New York, 1776–1777," *New York History*, January 1982, 42.

"Just one book had been published . . ." John S. Billings, "Literature and Institutions," in *A Century of American Medicine, 1776–1876*, Edward H. Clark et al. (Philadelphia: Henry C. Lea, 1876), 292.

"'The libraries of our physicians were composed . . .'" Billings, "Literature," 293–94.

"John Ranby, a British surgeon . . ." Blanco, "Military Medicine," 46.

"A quick-reference manual . . ." Lance P. Steahly and David W. Cannon Sr., Eds., *Evolution of Forward Surgery in the U.S. Army: From the Revolutionary War to the Combat Operations of the 21st Century* (Fort Sam Houston, TX: Borden Institute, 2018), 13.

"It is only necessary to remember . . ." Frederick S. Dennis, quoted in eulogy by John H. Bradshaw for James Rushmore Wood, reprinted in *Surgery, Gynecology, and Obstetrics*, March 1929, 441–45.

"Non-officers, though, would typically . . ." Elizabeth Rorke, "Surgeons and Butchers," *US History: Brandywine Battlefield Historic Site*, https://ushistory.org/brandywine. Access Aug. 19, 2024.

"He would clamp whatever large blood vessels . . ." "Revolutionary War Medicine, Battlefield Surgery, Amputation, and Smallpox Inoculation," www.mountvernon.org. Accessed August 22, 2024.

"Little information exists . . ." Alan J. Hawk, "ArtiFacts: Jean Louis Petit's Screw Tourniquet," *Clinical Orthopaedics and Related Research*, August 25, 2016, 2579.

"It demonstrated the need to create . . ." Derick Moore, "Fun Facts: From Counties Named Liberty to $368.6M Worth of Fireworks Sold," *US Census Bureau*, www.census.gov. Accessed August 22, 2024.

CHAPTER 7: RESURRECTION, DECAY, AND DISSECTION

"The group convened at the historic . . ." William H. Ukers, *All About Coffee* (New York: Project Gutenberg, 1922 [2009]), Chapter 10, 119.

"The board of governors for New York Hospital . . ." Eric Larrabee, *The Benevolent and Necessary Institution: The New York Hospital, 1771–1971* (Garden City, NY: Doubleday & Co., 1971), 88.

"Post was 'tall, handsome . . .'" Howard A. Kelly, *A Cyclopedia of American Medical Biography, Vol. 11* (Philadelphia: W. B. Saunders, 1912), 281.

"When he died in 1783 . . ." Melian Solly, "Why a London Museum Is Removing the Skeleton of an 'Irish Giant' from View," *Smithsonian Magazine*, www.smithsonianmag.com. Accessed September 17, 2024.

"On his death, said one newspaper . . ." John F. Fleetwood, "The Dublin Body Snatchers, Part 1," *Dublin Historical Record*, December 1998, 36.

"Instead his skeleton was placed on display . . ." The museum removed the skeleton from display in 2023 in response to "the sensitivities and the differing views surrounding the display and retention of Charles Byrne's skeleton." The skeleton will be retained for research purposes. "Statement on the Skeleton of Charles Byrne from the Board of Trustees of the Hunterian Collection," *Hunterian Museum*, https://hunterianmuseum.org. Accessed September 17, 2024.

"A student of Hunter's . . ." "Explore John Hunter's House," an online exhibit for the *Hunterian Museum*, https://hunterianmuseum.org. Accessed September 17, 2024.

"The ideal time for resurrecting a body . . ." Frederick C. Waite, "Grave Robbing in New England," *Bulletin of the Medical Library Association*, July 1945, 278.

"Author and historian James O. Breeden . . ." James O. Breeden, "Body Snatchers and Anatomy Professors: Medical Education in Nineteenth-Century Virginia," *The Virginia Magazine of History and Biography*, July 1975, 322.

"'No self-respecting grave robber . . .'" Suzanne M. Shultz, *Body Snatching: The Robbing of Graves for the Education of Physicians in Early Nineteenth Century America* (Jefferson, NC: McFarland, 2005), 32.

"'Neither the common nor the statutory law . . .'" Waite, "Grave Robbing," 272–73.

"'The amount of time required . . .'" James Blake Bailey, *The Diary of a Resurrectionist, 1811–1812: To Which Are Added an Account of the Resurrection Men in London and a Short History of the Passing of the Anatomy Act* (London: Swan Sonnenschein, 1896), 60.

"Certain country churchyards were selected . . ." Shultz, *Body Snatching*, 35.

"'I got the age, sex, race . . .'" Bill Traughber, "The Mystery of Col. William Shy's Burial in a Cast Iron Casket," *Williamson Herald*, April 26, 2023, www.williamsonherald.com. Accessed October 19, 2024.

"The University of Tennessee Anthropological Research Facility . . ." Mary Roach, *Stiff: The Curious Lives of Human Cadavers* (New York: W. W. Norton, 2003), 61.

"The grove is home to . . ." As of this writing, there are seven forensic anthropology facilities in the United States including the University of Tennessee, Knoxville: Colorado Mesa University, George Mason University, Northern Michigan University, Sam Houston State University, Texas State University, and Western Carolina University.

"The ARF holds 150–300 donated human remains . . ." Email from Dawnie Wolfe Steadman, director of the Forensic Anthropology Center, October 20, 2024.

"Researchers at the center . . ." Roach, *Stiff*, 62.

"Under normal conditions the body temperature . . ." Roach, *Stiff*, 61–62.

"The rapidity of postmortem cooling . . ." Abdulaziz M. Almulhim and Ritesh G. Menezes, "Evaluation of Postmortem Changes," [Updated 2023 May 1]. In: StatPearls [Internet]. Treasure Island (FL): StatPearls Publishing; January 2024.

Available from: https://www.ncbi.nlm.nih.gov/books/NBK554464. Accessed October 4, 2024.

"Blood pooling causes distinctive discoloration . . ." Almulhim, "Postmortem Changes."

"Lita Proctor, an National Institutes of Health scientist . . ." "NIH Human Microbiome Project Defines Normal Bacterial Makeup of the Body," National Institutes of Health, August 31, 2015, www.nih.gov. Accessed October 16, 2024.

"Explained world-renowned expert . . ." Roach, *Stiff*, 65.

"Occasionally, though, the abdominal wall . . ." Roach, *Stiff*, 68.

"Cadaverine's odor has been described . . ." "Molecule of the Week: Cadaverine and Putrescine," American Chemical Society, October 31, 2011, www.acs.org/molecule-of-the-week/archive. Accessed October 19, 2024.

"The rate of decay doubles . . ." Rutwik Shedge et al., "Postmortem Changes," StatPearls Publishing, January 2024, www.ncbi.nlm.nih.gov. Accessed October 19, 2024.

"Generally bodies older than . . ." Breeden, "Body Snatching," 324.

"'For example,' wrote physician John Shaw . . ." John Shaw, *A Manual for the Student of Anatomy: Containing Rules for Displaying the Structure of the Body, so as to Exhibit the Elementary Views of Anatomy and Their Application to Pathology and Surgery* (London: Burgers and Hill, 1821), vx–vxi.

"'They might have a knee one day . . .'" Ian Sample, "18th Century Doctors Shared Bodies to Teach Dissections, Research Shows," *Guardian*, February 15, 2015, www.theguardian.com. Accessed October 29, 2024.

"One student wrote in his notes . . ." Ralph C. Larrabee, "Student Notes on Anatomy," *Harvard Medical Library* (part of Ralph Clinton Larrabee papers, 1893–1933), https://collections.countway.harvard.edu. Accessed October 28, 2024.

"Instructor John Shaw used a particular formula . . ." Shaw, *Anatomy Manual*, 21.

"The procedure would be repeated . . ." Shaw, *Anatomy Manual*, 27.

"'The arteries may be injected . . .'" Shaw, *Anatomy Manual*, 33.

"Sometimes anatomists saved body parts . . ." Elizabeth T. Hurren, *Dissecting the Criminal Corpse Staging Post-Execution Punishment in Early Modern England* (London: Palgrave Macmillan, 2006), 122.

"Sometimes the flesh would be boiled down . . ." Hurren, *Dissecting the Criminal*, 42.

"Infants were often placed into coffins . . ." Jenna M. Dittmar and Piers D. Mitchell, "From Cradle to Grave via the Dissection Room: The Role of Foetal and Infant Bodies in Anatomical Education from the Late 1700s to Early 1900s," *Journal of Anatomy*, June 30, 2016, 721.

"Clossy taught in a room . . ." Thomas Gallagher, *A Doctor's Story: In Commemoration of the 200th Anniversary of the Columbia University College of Physicians and Surgeons* (New York: Harcourt, Brace & World, 1967), 24.

"Clossy then described the anatomy . . ." Samuel Clossy, *Observations on Some of the Diseases of the Parts of the Human Body. Chiefly Taken from the Dissections of Morbid Bodies* (London: G. Kearsley, 1763), 7–9.

CHAPTER 8: BUILDING A NATION

"The last British troops . . ." William A. Polf, *Garrison Town: The British Occupation of New York City, 1776–1783* (New York: New York State American Revolution Bicentennial Commission, 1976), 51.

"The fire destroyed Trinity Church . . ." Edwin G. Burrows and Mike Wallace. *Gotham: A History of New York City to 1898* (Oxford, UK: Oxford University Press, 2000), 265.

"He stayed only a brief time . . ." Mary Tsaltas-Ottomanelli, *Evacuation, Occupation, and Farewell: The Last Chapter of The Revolutionary War in New York City, Fraunces Tavern Museum*, www.frauncestavernmuseum.org/. Accessed August 28, 2024.

"Congress took up residence . . ." Burrows, *Gotham*, 266.

"It was the city's first Catholic church . . ." Burrows, *Gotham*, 273.

"Antifederalists argued that . . ." Ron Chernow, *Alexander Hamilton* (New York: Penguin Publishing Group, 2004), 244.

"Jay owned five slaves . . ." Burrows, *Gotham*, 285.

"Thomas Jefferson might have summed up . . ." "Thomas Jefferson to John Holmes, 22 April 1820," *Founders Online*, National Archives, https://founders.archives.gov/documents/Jefferson/03-15-02-0518. Accessed August 30, 2024. [Original source: The Papers of Thomas Jefferson, Retirement Series, vol. 15, 1 September 1819 to 31 May 1820, ed. J. Jefferson Looney (Princeton, NJ: Princeton University Press, 2018), 550–51.]

"'They were in deep financial difficulty . . .'" "Africans in America, Part 2, John Kaminski on Post-War America in the 1780s," *PBS Modern Voices Series*, www.pbs.org/wgbh. Accessed March 28, 2022.

"The troops were hired through a fund . . ." Joseph Parker Warren, "The Confederation and the Shays Rebellion," *American Historical Review*, October 1905, 43.

"Patrick Henry, perhaps the nation's . . ." Paul Aron, "Patrick Henry Smells a Rat," *American Heritage*, Summer 2017, www.americanheritage.com. Accessed September 2, 2024.

"Washington at one point decried . . ." Edward J. Larson, "Washington's Constitution," *George Washington's Mount Vernon*, www.mountvernon.org. Accessed September 2, 2024.

"Said Hamilton during the negotiations . . ." "New York Ratifying Convention Remarks (Francis Childs's Version)," *Flounders Online*, https://founders.archives.gov. Accessed September 4, 2024.

"Pierce Butler, a delegate from . . ." "Madison Debates," *Avalon Project, Yale Law School*, https://avalon.law.yale.edu. Accessed April 9, 2025.

"'A bill of rights is what the people . . .'" Thomas Jefferson, "From Thomas Jefferson to Uriah Forrest, with Enclosure, 31 December 1787," *Founders Online*, https://founders.archives.gov. Accessed September 1, 2024.

"He added, however, that 'the older I grow . . .'" Richard R. Beeman, "The Constitutional Convention of 1787: A Revolution in Government," *Constitution Center*, https://constitutioncenter.org. Accessed August 24, 2024.

"The Constitution was first printed . . ." John P. Kaminski et al. (Ed.), *The Documentary History of the Ratification of the Constitution Digital Edition* (Charlottesville, VA: University of Virginia Press, 2009), 131, https://archive.org. Accessed September 3, 2024.

"Federalists pointed out that . . ." Kaminski, *Documentary History*, 131.

"'The federalists evoked disunion . . .'" Chernow, *Hamilton*, 244.

"He enlisted Jay and Madison . . ." Hamilton invited Governor Morris to help, but he was too busy. William Duer, state senator of New York, wrote at least two papers, neither of which made Hamilton's final cut. Chernow, *Hamilton*, 248.

"In the introduction Publius urged citizens . . ." Alexander Hamilton, "The Federalist 1: To the People of the State of New-York," *Library of Congress*, https://guides.loc.gov. Accessed January 26, 2022.

"He told Washington in a February 1788 letter . . ." Walter Stahr, *John Jay: Founding Father* (New York: Diversion Books, 2012), 321.

"Hamilton and Madison churned out dozens . . ." Chernow, *Hamilton*, 249.

"Hamilton is said to have written . . ." "Federalist Papers: Primary Documents in American History," *Library of Congress*, https://guides.loc.gov. Accessed January 26, 2022.

"Ira C. Lupu, a nationally recognized scholar . . ." Ira C. Lupu, "The Most-Cited Federalist Papers," Constitutional Commentary, 418.

"Many small sections of the essays . . ." Alexander Hamilton, John Jay, and James Madison, *The Federalist: A Collection of Essays by Alexander Hamilton, John Jay, and James Madison* (New York: Wiley Book Co., 1901), 285.

"Simple, powerful words . . ." The Constitution was ratified first by Delaware, Pennsylvania, and New Jersey in December 1787, Georgia and Connecticut in January 1788, and Massachusetts in February. New York wouldn't ratify until July 26, 1788, by which point, with New Hampshire's vote, the document had become the law of the land.

"More than fifty years later . . ." National Park Service, "Arrival of the First Africans in 1619," www.nps.gov. Accessed November 3, 2024.

"There were eleven of them in all . . ." General Services Administration, *The New York African Burial Ground Unearthing the African Presence in Colonial New York* (Washington, DC: Howard University Press, 2009), 14.

"Between 1776 and 1800 more than 1.9 million people . . ." General Services Administration, *Unearthing*, 49.

"African residents soon began attending church . . ." "Beyond the Village and Back: Harlem's Mother A.M.E. Zion Church," *Village Preservation*, www.village preservation. Accessed September 9, 2024.

"Members of a Dutch Reformed church in Brooklyn . . ." Andrea C. Mosterman, "The Forgotten History of Slavery in New York," *Cornell University Press*, www .cornellpress.cornell.edu. Accessed September 11, 2024.

"Malnutrition among enslaved . . ." General Services Administration, *Unearthing*, 132.

"White citizens, even though exposed to the same . . ." General Services Administration, *Unearthing*, 186.

"The Common held a tannery . . ." Statistical Research, Inc., *The Skeletal Biology, Archaeology, and History of the New York African Burial Ground: A Synthesis of Volumes 1, 2, and 3* (Washington, DC: Howard University Press, 2009), 56.

"Eighteenth-century historian David Valentine . . ." Edna G. Medford, "Historical Perspectives of the African Burial Ground New York: Blacks and the Diaspora," *Howard University Press in Association with the General Services Administration*, 2009, 2.

"The area was labeled on a 1755 map . . ." "African Burial Ground," *National Park Service History eLibrary*, https://npshistory.com. Accessed September 13, 2024.

"However, a small section of the burial ground . . ." Caroline de Costa and Francesca Miller, "American Resurrection and the 1788 New York Doctors' Riot," *Lancet*, January 22, 2011, 293.

"Historians believe those individuals . . ." Statistical Research, *Skeletal Biology*, 224.

"Nearly all bodies were buried . . ." Statistical Research, *Skeletal Biology*, 71.

"Clam and oyster shells were sometimes included . . ." Statistical Research, *Skeletal Biology*, 234.

"'It was the place where kin were buried . . .'" General Services Administration, *Unearthing*, 259.

CHAPTER 9: TENSIONS SURGE

"Newspapers accused Clossy . . ." Russell W. Irvine, "Pride and Prejudice," *Journal of the College of Physicians and Surgeons*, Winter 2000, www.cuimc.columbia.edu. Accessed July 12, 2024.

"Classes resumed with nine professors . . ." *A History of Columbia University, 1754–1904: Published in Commemoration of the One Hundred and Fiftieth Anniversary of the Founding of King's College* (New York: Columbia University Press, 1904), 312.

"McKnight was a surgeon . . ." Johann Hermann Baas and Henry E. Handerson, *Outlines of the History of Medicine and the Medical Profession* (New York: J. H. Vail, 1889), https://archive.org, 817. Accessed July 12, 2024.

"He served as a chief hospital physician . . ." Thomas Gallagher, *A Doctor's Story: In Commemoration of the 200th Anniversary of the Columbia University College of Physicians and Surgeons* (New York: Harcourt, Brace & World, 1967), 49.

"One citizen, writing to the editor ..." Unknown author, *Daily Advertiser*, April 23, 1788, 2.

"For instance, a newspaper in 1763 ..." Jules Calvin Ladenheim, "'The Doctors' Mob' of 1788," *Journal of the History of Medicine*, Winter 1950, 25.

"He claimed that graveyards ..." Robert J. Swan, "Prelude and Aftermath of the Doctors' Riot of 1788: A Religious Interpretation of White and Black Reaction to Grave Robbing," *New York History*, October 2000, 428–29.

"A third writer, A Physician ..." Swan, "Prelude," 429.

"'New York had not yet ratified ...'" Swan, "Prelude," 429.

"Black church patrons were buried ..." Taken from vestry minutes from April 12, 1790, per email with Kathryn Hurwitz, Assistant Archivist at Trinity Church Wall Street, September 1, 2022.

"Swan explained that 'interment ...'" Swan, "Prelude," 427.

"On February 3, 1788, a petition arrived ..." Swan, "Prelude," 438.

"The "petition deferential and ..." Caroline de Costa and Francesca Miller, "American Resurrection and the 1788 New York Doctors' Riot," *Lancet*, January 22, 2011, 293.

"The mayor, James Duane ..." Robert J. Swan, "Prelude and Aftermath of the Doctors' Riot of 1788: A Religious Interpretation of White and Black Reaction to Grave Robbing," *New York History*, October 2000, 419.

"The "document was ..." Swan, "Prelude," 440.

"By 1788 a total of sixteen churches ..." Swan, "Prelude," 439.

"Church leaders responded ..." Trinity Church Vestry Minutes from February 20, 1788, per email from Kathryn Hurwitz, Associate Archivist at Trinity-Wall Street Church, August 31, 2022.

"Thomas Gallagher, in his history ..." Gallagher, *Story*, 47.

"'Mr. Printer,' the letter began ..." *Daily Advertiser*, February 16, 1788, 3.

"Six days later the newspaper ..." Gallagher, *Story*, 58.

"The writer signed his screed ..." Gallagher, *Story*, 57–58.

"They banged on the man's door ..." Gallagher, *Story*, 61.

"Student of Physic denied any shame ..." Swan, "Prelude," 446.

"Directing his note to the paper's publisher ..." Gallagher, *Story*, 61.

"Philo-Smilio, with tongue planted firmly ..." Gallagher, *Story*, 60.

"The writer then added ...'" Swan, "Prelude," 447. The full quote, from Job 3:17–19 in the King James Version of the Bible, "There the wicked cease from troubling; And there the weary be at rest."

"By the end of February ..." Swan, "Prelude," 438.

"I shall not fill up your paper ..." Gallagher, *Story*, 62.

"'By 1788, interred there ...'" Swan, *Prelude*, 443.

Besides Hicks, Bayley taught another student of note in 1788, a young man named David Hosack, who would go on to treat Alexander Hamilton's son, Philip, when he developed yellow fever in the late 1790s. Philip, just fifteen at the time, would become seriously ill from the infection, finally slipping into a coma. Hosack would save him by immersing the boy in a bath of rum and Peruvian bark, which contains quinine, a common treatment for yellow fever and malaria. Tragically, Philip lived just another four years, succumbing to a bullet wound received in a duel with a lawyer who had called Philip and a friend "damned rascals." Hosack could do nothing to save Philip's life that time, nor would he be able to save his famous father's life three years later, when he was fatally wounded by Aaron Burr, dueling on the same grounds in Weehawken, New Jersey, where his son had been shot. But in early 1788, nineteen-year-old David Hosack and twenty-year-old John Hicks Jr. were both studying anatomy on their way to becoming physicians.

" Swan said that Hicks . . ." Gallagher, *Story,* 65.

"Usually, the students had contented . . ." Joel Tyler Headley, *The Great Riots of New York, 1712–1873: Including a Full and Complete Account of the Four Days' Draft Riot of 1863* (New York: E. B. Treat, 1873), 57.

"At the time the city was home . . ." "Early American Newspapers, 1780–1789," *Readex,* www.readex.com. Accessed November 9, 2024.

"That act, commonplace though it was . . ." Paul A. Gilje, *Rioting in America* (Bloomington, IN: Indiana University Press, 1996), 51.

CHAPTER 10: THE OBNOXIOUS DR. HICKS

"Governor George Clinton was there . . ." Thomas Gallagher, *A Doctor's Story: In Commemoration of the 200th Anniversary of the Columbia University College of Physicians and Surgeons* (New York: Harcourt, Brace & World, 1967), 64.

"Bayley's anatomy lab was located . . ." Gallagher, *Doctor's Story,* 64.

"The boy's father was a Freemason . . ." Swan, "Prelude," 447.

"Bayley and his students were working on . . ." Whitfield J. Bell Jr., "Doctors' Riot, New York, 1788," *Bulletin of the New York Academy of Medicine,* December 1971, 1501–2.

"One hid in a chimney . . ." Bell, "Doctors' Riot," 1502.

"The men soon discovered a third body . . ." Bell, "Doctors' Riot," 1502.

"An article in Boston's . . ." Editorial by Order, Decency, Justice, and Peace, *American Herald,* May 5, 1788, 2. Reprinted from *Daily Patriotic Register,* April 19, 1788.

"The rowdies beat them all . . ." Swan, "Prelude," 448.

"As luck would have it the city's mayor . . ." Scott McCabe, "Crime History: 'Rattle Watch' Becomes Original New World Police Force," *Washington Examiner,* August 11, 2012, www.washingtonexaminer.com. Accessed November 21, 2024.

"The city lacked any kind of organized . . ." "The History of New York City Police Department," *National Criminal Justice Reference Service,* 1993, www.ojp.gov/ncjrs. Accessed November 21, 2024.

"Some of the men carried the remains ..." Jules Calvin Ladenheim, "'The Doctors' Mob' of 1788," *Journal of the History of Medicine,* Winter 1950, 30.

CHAPTER 11: HUNTING FOR THE CURSED DISSECTORS

"Although Duane lived well outside downtown ..." Bowery Boys, "Meet James Duane, New York's First Mayor After the American Revolution," www.boweryboys history.com/. Accessed June 30, 2022.

"Jay, serving at the time ..." George Pellew, *John Jay: American Statesman* (Cambridge, MA: Houghton, Mifflin, and Co., 1898), 255.

"Jay had built a 'large, square ...'" Walter Starr, *John Jay: Founding Father* (New York: Diversion Books, 2012), 223.

"Clinton lived at the Governor's Mansion ..." "George Clinton Historical Marker," Historical Marker Database, www.hmdb.org/. Accessed October 22, 2024. The house served as the Governor's Mansion from 1784 to 1791. Washington used the house as headquarters when he took command of the troops in 1776.

"Two men, a medical student named ..." Robert J. Swan, "Prelude and Aftermath of the Doctors' Riot of 1788: A Religious Interpretation of White and Black Reaction to Grave Robbing," *New York History,* October 2000, 448.

"'If the indictment was based on ...'" Ralph G. Victor, "An Indictment for Grave Robbing in New York, Doctors' Riot, 1788," *Annals of Medical History*, September 1940, 366–70.

"Said Victor, 'No New York newspaper ...'" Victor, "Indictment," 369.

"The rioters pushed their way through ..." Thomas Gallagher, *A Doctor's Story: In Commemoration of the 200th Anniversary of the Columbia University College of Physicians and Surgeons* (New York: Harcourt, Brace & World, 1967), 71.

"Duane addressed the crowd and ..." Jules Calvin Ladenheim, "'The Doctors' Mob' of 1788," *Journal of the History of Medicine,* Winter 1950, 31.

"Cochran let a small group of rioters ..." Ladenheim, "Mob," 31.

"Similar scenes played out ..." Swan, "Prelude," 448.

"They 'swarmed without opposition ...'" Swan, "Prelude," 448.

"Published the following day ..." Notice in the *Daily Advertiser*, April 15, 1788, 2.

"Even more outlandish ..." Notice in the *Daily Advertiser*, April 16, 1788, 2.

"Leaders at the front of the mob ..." Joel Tyler Headley, *The Great Riots of New York, 1712–1873: Including a Full and Complete Account of the Four Days' Draft Riot of 1863* (New York: E. B. Treat, 1873), 62.

"An imposing stone building ..." Tom Miller, "The Lost 1756 Debtors' Prison–City Hall Park," https://daytoninmanhattan.blogspot.com. Accessed November 22, 2024.

"The troops broke down the door ..." Scott M. Smith, "Major Robert Rogers and the American Revolution," *Journal of the American Revolution*, November 23, 2021.

"The crowd increased throughout the day ..." Swan, "Prelude," 448.

"He asked for volunteers ..." Swan, "Prelude," 450.

"Without orders to use force ..." "Boston April 28," *Boston Gazette*, April 28, 1788, 2.

"With doors several inches thick ..." Phone conversation with Tom Hennessy, curator of the Lock Museum of America in Terryville, Pennsylvania, December 3, 2024.

"'Every picket fence within blocks of the jail ...'" Ladenheim, "Mob," 33.

"Some protesters, however ..." "New-York: April 15," *New-York Packet*, April 15, 1788, 2.

"'Throughout, the rioters maintained a sense of purpose ...'" Paul A. Gilje, *The Road to Mobocracy: Popular Disorder in New York City, 1763–1834* (Chapel Hill, NC: Omohundro Institute of Early American History and Culture and the University of North Carolina Press, 1988), 80.

"As they made their way along ..." "New-York: April 25," *New-York Packet*, April 25, 1788, 2.

CHAPTER 12: BATTLE OF BRIDEWELL

"Mayor Duane had been scurrying about ..." Whitfield J. Bell Jr., "Doctors' Riot, New York, 1788," *Bulletin of the New York Academy of Medicine*, December 1971, 1503.

"Clarkson was Brigadier General ..." "General Matthew Clarkson: A Biographical Sketch," *New York Heritage Digital Collection*, https://nyheritage.contentdm.oclc .org. Accessed December 6, 2024.

"After sending the last cohort ..." Robert J. Swan, "Prelude and Aftermath of the Doctors' Riot of 1788: A Religious Interpretation of White and Black Reaction to Grave Robbing," *New York History,* October 2000, 450.

"'My God, Jay!'" Thomas Gallagher, *A Doctor's Story: In Commemoration of the 200th Anniversary of the Columbia University College of Physicians and Surgeons* (New York: Harcourt, Brace & World, 1967), 80–81.

"'Handing Clarkson one sword ...'" Walter Starr, *John Jay: Founding Father* (New York: Diversion Books, 2012), 253.

"Also present was General John Lamb ..." Swan, "Prelude," 450.

"... who had commanded artillery troops ..." "Colonel John Lamb," *National Park Service*, https://nps.gov. Accessed December 7, 2024.

"Clarkson's daughter, Mary Rutherfurd ..." "Peter Augustus Jay and Mary Rutherfurd Clarkson," *Jay Heritage Center*, https://jayheritagecenter.org. Accessed December 6, 2024.

"At the door now were a handful ..." Jules Calvin Ladenheim, "'The Doctors' Mob' of 1788," *Journal of the History of Medicine,* Winter 1950, 30.

"'We marched up to the jail ...'" Unknown letter writer, "Boston, April 28," *Boston Gazette*, 2.

The city's population swelled in the late 1700s and early 1800s, so the city began to expand northward, endangering the African Burial Ground and potter's field. The city established, in 1794, a public burial ground at the corner of Fifth Avenue and Broadway at 23rd Street, land currently covered by Madison Square. The bodies there were moved three years later to the land now holding Washington Square Park, then to Bryant Square, and then to the corner of Third Avenue and 50th Street. The city's potter's field is now located on Hart's Island in the East River. Source: Charles D. Cheek and Daniel G. Roberts, *The Archeology of 290 Broadway, Volume I: The Secular Use of Lower Manhattan's African Burial Ground*, Report prepared for Jacobs Edwards and Kelcey and the US General Services Administration, 2009, 133.

"Robert Swan has outlined his reasons . . ." Swan, "Prelude," 452.

"Newspaper articles from the time . . ." "New-York: April 15," *New-York Packet*, April 15, 1788, 2.

"'The interments not only of strangers . . .'" "New-York: April 25," *New-York Packet*, April 25, 1788, 2.

"Just then about 100 troops . . ." Unknown, "Boston," 2.

"Along with Armstrong's cohort . . ." Bell, "Doctors' Riot," 1501.

"Duane was struck in the head . . ." Jules Calvin Ladenheim, "'The Doctors' Mob' of 1788," *Journal of the History of Medicine*, Winter 1950, 34.

"He had never been a fan of mobs . . ." Paul Douglas Lockhart, *The Drillmaster of Valley Forge: The Baron de Steuben and the Making of the American Army* (New York: HarperCollins ebooks, 2008), 293.

"However, when a brickbat struck Steuben . . ." Bell, "Doctors' Riot," 1503.

"whatever benevolence he felt . . ." Ladenheim, "Doctors' Mob," 34.

"'We drove them with our bayonets . . .'" Thomas Gallagher, *A Doctor's Story: In Commemoration of the 200th Anniversary of the Columbia University College of Physicians and Surgeons* (New York: Harcourt, Brace & World, 1967), 82.

"The crowd then stormed around . . ." Unknown, "Boston," 2.

"Dozens of stones, planks . . ." Bell, "Doctors' Riot," 1503.

"The family doctor, wrote Sarah . . ." Gallagher, *Doctors' Story*, 84.

"Jay survived both bodily insults . . ." Walter Stahr, *John Jay: Founding Father* (New York: Diversion Books, 2012), 253.

"The next morning, Tuesday . . ." Swan, "Prelude," 450.

"Malcolm was a Scot who had immigrated . . ." Katharine Schuyler Baxter, *A Godchild of Washington, a Picture of the Past* (New York: F. Tennyson Neely, 1897), 440.

"Bauman had started the war as a captain . . ." "To Thomas Jefferson from Sebastian Bauman, 29 May 1797," *Founders Online*, https://founders.archives.gov. Accessed December 18, 2024.

"To avoid capture by the rioters . . ." Ladenheim, "Doctors' Mob," 35.

"Protesters mistakenly attacked . . ." Ladenheim, "Doctors' Mob," 35.

"He was actually a diplomat . . ." "Sir John Temple," Massachusetts Historical Society, www.masshist.org. Accessed December 19, 2024.

"No mob collected again . . ." *New-York* Packet April 25, 2.

". . . in a two-story building . . ." "Charles Wiesenthal and the Original Medical Society," *MedChi Archives*, https://medchiarchives.blogspot.com. Accessed December 30, 2024.

"The body of an executed murderer . . ." Suzanne M. Shultz, *Body Snatching: The Robbing of Graves for the Education of Physicians in Early Nineteenth Century America* (Jefferson, NC: McFarland, 2005), 46.

"The building, located at the corner . . ." Shultz, *Body Snatching*, 46.

"'We know little about this ill-fated lady . . .'" Hannibal Hamlin, "The Dissection Riot of 1824 and the Connecticut Anatomical Law," *Yale Journal of Biology and Medicine*, March 1935, 280.

"'It was opened . . .'" Hamlin, "Riot of 1824," 278.

"He must have been heartstruck . . ." Hamlin, "Riot of 1824," 278.

"'A drum has beat . . .'" Hamlin, "Riot of 1824," 279.

"Luckily he recognized the threat and ran . . ." Hamlin, "Riot of 1824," 284.

"He was convicted by a jury . . ." Hamlin, "Riot of 1824," 282.

"An article in the *Connecticut Herald* . . ." Erik Ofgang, "When Yale Medical Students Robbed a Grave for Science, New Haven Erupted in Fury," *CTInsider.com*, March 19, 2028, www.ctinsider.com. Accessed January 31, 2025.

"Michael Sappol, author of *A Traffic* . . ." Ofgang, "New Haven."

"Fenner had been telling people . . ." Ben C. Clough, "The Corpse and the Beaver Hat," *Brown Alumni Monthly*, May 1961, https://archive.org. Accessed April 10, 2025.

"'He not only told the story . . .'" Clough, "Beaver Hat," 10.

"He must have hated the judge . . ." Clough, "Beaver Hat," 13.

"'Although Judge Dorrance was of good character . . .'" Shultz, *Body Snatching*, 48–49.

CHAPTER 13: BURKING, BONE BILLS, AND EMBALMING

"He tried his hand . . ." George MacGregor, *The History of Burke and Hare and of the Resurrectionist Times: A Fragment from the Criminal Annals of Scotland* (London: Hamilton, Adams, and Co., 1884), 47.

"He joined the militia at age . . ." William Roughead, Ed., *Burke and Hare* (London: William Hodge and Co., 1921), 12.

"He was described as . . ." Roughead, *Burke*, 14.

"Hare was born outside of Sarva . . ." "William Hare," *Murderpedia*, https://murderpedia.org. Accessed January 5, 2025.

"One of Knox's assistants . . ." Roughead, *Burke*, 15.

"The bill stated clearly that medical students . . ." "Anatomy Act of 1832," *Irish Statute Book*, www.irishstatutebook.ie. Accessed January 12, 2025.

"Most important, though, the act expanded . . ." Anatomy Act, *Irish Statute*.

"Nineteen more medical schools . . ." Robert G. Slawson, "Medical Training in the United States Prior to the Civil War," *Journal of Evidence-Based Complementary & Alternative Medicine*, January 3, 2012, DOI: 10.1177/2156587211427404. Accessed December 22, 2025.

"New York led the list with . . ." Bess Lovejoy, "Body Snatchers of Old New York," *Lapham's Quarterly*, October 31, 2013, www.laphamsquarterly.org. Accessed January 1, 2025.

"Anyone convicted of it . . ." "An Act to Prevent the Odious Practice of Digging Up and Removing for the Purpose of Dissection, Dead Bodies Interred in Cemeteries or Burial Places," *New York State Law, 12th Section, Chapter 3*. Passed January 6, 1789.

"Perpetrators could be fined . . ." Frederick C. Waite, "Grave Robbing in New England," *Bulletin of the Medical Library Association*, July 1945, 274.

"Anatomists receiving a body were . . ." Massachusetts. General Court. House of Representatives. Committee on the Judiciary, 1834. *Center for the History of Medicine* (Francis A. Countway Library of Medicine), https://collections.countway.harvard.edu. Accessed January 10, 2025.

"'The vast majority of illegal disinterments . . .'" Waite, "Grave Robbing," 275.

"Numerous other medical colleges . . ." Claude Heaton, "Body Snatching in New York City," *New York State Journal of Medicine*, October 1, 1943, 1861–1865.

"In the four-year period . . ." Mathis, *Antebellum America*.

". . . transported to Augusta in old whiskey . . ." "'Resurrection Man' Supplied Cadavers to Medical College," *Augusta Chronicle*, August 29, 2010, www.augustachronicle.com. Accessed January 2, 2025.

"Grandison Harris, purchased for $700 . . ." "Known but to God: The Story of the Resurrection Man," *Augusta University Jagwire*, https://jagwire.augusta.edu. Accessed January 2, 2025.

"Undertakers during the Civil War tried . . ." Erich Brenner, "Human Body Preservation: Old and New Techniques," *Journal of Anatomy*, September 2014, 319.

"After hearing the news and composing himself . . ." "The Sons: Elmer Ellsworth (1837–1861)," *Mr. Lincoln and Friends.org*, www.mrlincolnandfriends.org. Accessed January 16, 2025.

"The pump was turned on . . ." "A Wisconsin Civil War Story," *Wisconsin History*, https://wisconsinhistory.org. Accessed January 16, 2025.

"At Lincoln's direction Holmes set up a team . . ." Susan Parsons, "Preserving the Union: A History of Modern Embalming," www.cayugacounty.us/DocumentCenter/View/1765/Embalming-PDF. Accessed January 16, 2025.

"...including eight generals ..." Robert G. Mayer, *Embalming: History, Theory, and Practice, Fifth Edition* (New York: McGraw-Hill, 2012), 488.

"Holmes, who would earn the nickname ..." "Embalming and the Civil War," *National Museum of Civil War Medicine*, February 20, 2016, www.civilwarmed.org. Accessed January 16, 2025. Only Union troops were embalmed. Southern states lacked morticians with a knowledge of embalming, and even if they had the knowledge, the solutions used were manufactured only in the North.

"...out of roughly 698,000 ..." Joan Barcelóa et al., "New Estimates of US Civil War Mortality from Full-Census Records," *Proceedings of the National Academy of Sciences*, November 18, 2024, www.pnas.org. Accessed January 16, 2025.

"Scientists believe that formaldehyde ..." Luoping Zhang, *Formaldehyde: Exposure, Toxicity and Health Effects* (London: Royal Society of Chemistry, 2018), 2.

"The gas is released in tobacco smoke ..." Jessica Clifton, "What Is Formaldehyde," *ReAgent Chemical Services*, www.chemicals.co.uk. Accessed February 1, 2025.

"'Cadavers obtained from legal sources ...'" Suzanne M. Shultz, *Body Snatching: The Robbing of Graves for the Education of Physicians in Early Nineteenth Century America* (Jefferson, NC: McFarland, 2005), 90.

"The organized arm of medicine ..." Schultz, *Robbing*, 91.

"By 1954 medical schools existed ..." Shultz, *Robbing*, 93.

"'For a dying man it is not a difficult decision ...'" Christiaan Barnard, *One Life* (Toronto: Macmillan, 1969), 261–62.

"The Uniform Law Commission ..." Robert A. Stein, *Forming a More Perfect Union: A History of the Uniform Law Commission* (Charlottesville: Matthew Bender, 2013), 12.

"The document, called the Uniform Anatomical Gift Act ..." Ghosh, "Human Cadaveric Dissection," 163.

"'We're stewards of your body ...'" Michael DePeau-Wilson, "Training on Cadavers Still Essential for Medical Students," *MedPage Today*, September 23, 2024, www.medpagetoday.com. Accessed March 3, 2025.

"Body brokers obtain most cadavers ..." Jeong Suh, "The Body Trade: Cashing In on the Donated Dead," *Reuters*, October 24, 2017.

"An investigative report by Reuters in 2017 ..." Suh, "Cashing In."

"Tanya Marsh, a law professor at ..." Mike Hixenbaugh et al., "Dealing Corpses from a Las Vegas Strip Mall: A Look Inside the Shadowy U.S. Body Trade," *NBC News*, December 11, 2024, www.nbcnews.com. Accessed February 10, 2025.

"'It's a wild, wild West out there ...'" "Thriving Trade in Body Parts Under Scrutiny," *NBC News*, March 7, 2004, www.nbcnews.com. Accessed February 10, 2025.

"Mastromarino said simply ..." Randall Patterson, "The Organ Grinder," *New York Magazine*, October 6, 2006, www.nymag.com. Accessed January 15, 2017.

"In one, a disgraced chiropractor ...", "Dealing Corpses from a Las Vegas Strip Mall: A Look Inside the Shadowy U.S. Body Trade," *NBC News Special Report:*

Dealing the Dead, December 11, 2024, www.nbcnews.com. Accessed February 13, 2025.

"He sold five sets of arms . . ." NBCNews.com, "Dealing Corpses."

"A survey in 2023 . . ." Eli Shupe, Serena Karim, and Daniel Sledge, "Unclaimed Bodies and Medical Education in Texas," *Journal of the American Medical Association*, August 24, 2023, 1189.

"Patman told NBC that Honey . . ." Cara Lynn Shultz, "Veteran's Family Says His Body Was Cut Up, Sold Without Consent," *People.com*, September 17, 2024, https://people.com. Accessed March 18, 2025.

"Honey's body was dismembered at the UNT program . . ." Mike Hixenbaugh, Jon Schuppe, and Susan Carroll, "Cut Up and Leased Out, the Bodies of the Poor Suffer a Final Indignity in Texas," *NBC News Special Report: Dealing the Dead*, September 16, 2024, www.nbcnews.com. Accessed March 17, 2025.

"A medical education company in Pittsburgh . . ." Hixenbaugh, "Veteran's Corpse."

"The university suspended . . ." Andy North, "Important Update on Willed Body Program," *University of North Texas Health Sciences Center at Fort Worth*, September 13, 2024, www.unthsc.edu. Accessed February 13, 2025.

"Pauley used the payment application . . ." Dale Ellis, "Little Rock Woman Who Admitted to Selling Body Parts from UAMS Cadavers Sentenced to 15 Years in Prison," *Arkansas Online*, www.arkansasonline.com, January 17, 2025. Accessed February 18, 2025.

"Scott also sold the bodies . . ." Keith Schweigert, "Member of Human Remains Trafficking Conspiracy Will Serve 15 Months in Prison after Pleading Guilty," *Fox43*, www.fox43.com, January 23, 2025. Accessed February 18, 2025.

"Pauley pleaded guilty to state charges . . ." K. C. Baker, "Man Sentenced After Police Found Buckets of Human Remains in Harvard Medical School Scandal," *People.com*, March 5, 2024.

"'We want our surgeons to practice before . . .'" CBS, *Body Brokers*. Original quote edited for clarity and approved by Agent Johnson on March 6, 2025.

"Angela McArthur, director of the Anatomy Bequest Program . . ." Suh, "Cashing In."

"'Right now there are things going on . . .'" Phone interview with Thomas Champney, March 3, 2025.

"'It has very robust standards . . .'" Phone interview with Angela McArthur, March 7, 2025.

"'It's a major life decision . . .'" Champney phone interview.

"Donors should ask . . ." Jeffrey A. Zealley et al., "Human Body Donation: How Informed Are the Donors?" *Clinical Anatomy*, August 25, 2021, 19–25, https://doi.org/10.1002/ca.23780. Accessed April 1, 2025.

"Among the most common pieces of misinformation . . ." Interview with Paul Micah Johnson, March 25, 2025.

"'My father has been gone for over a decade . . .'" Marie Holmes, "My Loved One Donated Their Body to Science: This Is What It Meant for My Grief," *Huffpost.com*, April 24, 2023, www.huffpost.com. Accessed March 31, 2025.

"'If someone wants their body used only . . .'" Johnson phone interview.

"Jackie Dent, author of *The Great Dead Body Teachers* . . ." Jackie Dent, "My Grand-parents Donated Their Bodies to Science; I Needed to Know What Happens After," *Guardian*, March 3, 2023, www.theguardian.com. Accessed March 31, 2025.

"'I couldn't afford . . .'" Brian Grow and John Shiffman, "The Body Trade: A Reuters Journalist Bought Human Body Parts, Then Learned a Donor's Heart-Wrenching Story," *Reuters*, October 25, 2017.

"'I haven't seen anything this egregious . . .'" Grow, "Body Trade."

CHAPTER 14: ULYSSES, SIR, AND CELEBRATING DONORS

"'I'll never forget it . . .'" Phone interview with Dawn Lynch, June 13, 2024.

"'The evidence of the disease startled us . . .'" Howard Chang, "The Human Behind the Body: A Medical Student's Experience with Cadaveric Dissection," *Johns Hopkins Medicine: Biomedical Odyssey*, https://biomedicalodyssey.blogs .hopkinsmedicine.org. Accessed March 2, 2025.

"'The scariest part of anatomy . . .'" Chang, "Cadaveric Dissection."

"'Over time,' said another student . . ." Chang, "Cadaveric Dissection."

"'Medical students learn to deny . . .'" Pauline W. Chen, *Final Exam: A Surgeon's Reflections on Morality* (New York: Alfred A. Knopf, 2007), 17.

"Sindhu Ragunathan, a first-year medical student . . ." Bridjes O'Neil, "SLU Medical Students Honor 'Invaluable' Gift of Body Donors with Memorial Service," *Saint Louis University*, October 29, 2024, https://www.slu.edu/news. Accessed March 23, 2025.

"'I found I could not effectively work . . .'" Chen, *Final Exam*, 15–16.

"'I found I could not effectively work . . .'" Madeleine Spencer, "Life, Learning, and Dissection: Reflections on My Time in a Medical School Cadaver Lab," *MaddieMonster.com*, https://maddiemonster.artstation.com. Accessed March 23, 2025.

"'I had studied anatomy books for years . . .'" Spencer, "Reflections."

"'The table uses highly-detailed images . . ." Eirini-Maria Kavvadia et al., "The Anatomage Table: A Promising Alternative in Anatomy Education," *Cureus Journal of Medical Science*, August 6, 2023, DOI: 10.7759/cureus.43047. Accessed April 2, 2025.

"'The software allows students . . ." "Revolutionizing Medical Education: The HoloAnatomy Experience," *Texas A&M Technology Services*, May 2024, https://it.tamu.edu. Accessed April 2, 2025.

"At this writing a single . . ." "Anatomage Table 8.0 is a True Game Changer for Students," *Michigan Tech Biological Sciences Blog*, https://blogs.mtu.edu/biological. Accessed April 3, 2025.

". . . and in 2024 one US university . . ." AlensiaXR sales sheet for Florida Atlantic University; proposal accepted in 2024, https://techfee.fau.edu/approvedproposals/Download.cfm?sid=4727&pid=3215. Accessed April 3, 2025.

"One study during the COVID-19 pandemic . . ." Natalia Sinou et al., "Virtual Reality and Augmented Reality in Anatomy Education During COVID-19 Pandemic," *Cureus Journal of Medical Science*, February 19, 2023, DOI 10.7759/cureus.35170. Accessed March 31, 2025.

"'We're told the donor is your first patient . . .'" Zelda Blair, "Reflection," unpublished essay from University of Minnesota Medical School, Twin Cities Campus, https://med.umn.edu. Accessed March 20, 2025.

"We hope you realize what a lasting impact . . ." Unknown anatomy graduate student, "Letters of Appreciation," *Ohio State University*, https://medicine.osu.edu. Accessed March 23, 2025.

"'We have a huge responsibility . . .'" Bryan Luhn, "Reflections on Life, Death and Humanity," *University of Houston*, https://stories.uh.edu. Accessed March 20, 2025.

"A vase filled with white roses . . ." "PCOM Reflects During Celebration of Remembrance," *PCOM*, www.pcom.edu, May 4, 2022. Accessed April 4, 2025.

"I do not know how I could ever repay . . ." Samantha Cooper, "Reading for 2022 Celebration of Remembrance," *Pennsylvania College of Osteopathic Medicine*, https://www.pcom.edu. Accessed April 4, 2025.

ACKNOWLEDGMENTS

"As an investigator for the Detroit office . . ." Justin Sherman, "Inside the Largely Unregulated Market for Bodies Donated to Science: 'It's Harder to Sell Hot Dogs on a Cart,'" *CBS Reports*, March 23, 2023, www.cbsnews.com. Accessed April 11, 2025.

SELECTED REFERENCES

Bell, Whitfield J. 1965. *John Morgan: Continental Doctor*, University of Pennsylvania Press, Philadelphia.

Chernow, Ron. 2004. *Alexander Hamilton*, Penguin Books, New York.

Corner, Betsy Copping. *William Shippen, Jr: Pioneer in American Medical Education*, American Philosophical Society, Philadelphia.

Duer, William Alexander. 1849. *New-York as It Was During the Latter Part of the Last Century: An Anniversary Address Delivered Before the St. Nicholas Society of the City of New-York*. New York: Stanford and Swords.

Gallagher, Thomas. 1967. *The Doctor's Story: In Commemoration of the 200th Anniversary of the Columbia University College of Physicians and Surgeons*, Harcourt, Brace & World, New York.

Gilje, Paul A. 1996. *Rioting in America*, Indiana University Press, Bloomington.

Ladenheim, Jules Calvin. "The Doctors' Mob of 1788," *Journal of the History of Medicine*, Winter 1950.

Larrabee, Eric. 1971. *The Benevolent and Necessary Institution: The New York Hospital, 1771–1971*, Doubleday & Company, New York.

Richardson, Ruth. 2000. *Death, Dissection and the Destitute*, University of Chicago Press.

Roach, Mary. 2003. *Stiff: The Curious Lives of Human Cadavers*, W. W. Norton and Co., New York.

Shultz, Suzanne M. 1992. *Body Snatching: The Robbing of Graves for the Education of Physicians in Nineteenth Century America*, McFarland & Company, Jefferson, North Carolina.

Stahr, Walter. 2012. *John Jay: Founding Father*, Diversion Books, New York.

Stookey, Byron. 1962. *A History of Colonial Medical Education*, Springfield, Illinois.

Swan, Robert. 2000. "Prelude and Aftermath of the Doctors' Riot of 1788: A Religious Interpretation of White and Black Reaction to Grave Robbing," *New York History,* August, 417–56.

Walsh, James J. 1919. *History of Medicine in New York: Three Centuries of Medical Progress*, Vol. 4, National American Society, New York.

INDEX

ABOUT THE AUTHOR

ANDY MCPHEE IS A FORMER REGISTERED NURSE, NURSE EDUCATOR, and retired educational healthcare publisher. He is a two-time winner of the Association of Educational Publishers' award for feature writing and has written four books and hundreds of health and science articles for young adults. He is the author of *Donora Death Fog: Clean Air and the Tragedy of a Pennsylvania Mill Town*, published by The University of Pittsburgh Press in 2023. He lives with his wife and two dogs in Saint Petersburg, Florida.

www.ingramcontent.com/pod-product-compliance
Lightning Source LLC
Chambersburg PA
CBHW041831090426
42811CB00047B/2462/J